CW01666984

Yamaha YB 100 Owners Workshop Manual

by Pete Shoemark

Models covered:

YB 100 97 cc introduced UK May 1973
YB 100E 97 cc introduced UK 1979

ISBN 0 85696 474 3

© Haynes Publishing Group 1979

All rights reserved. No part of this book may be reproduced or transmitted in any form or by any means, electronic or mechanical, including photocopying, recording or by any information storage or retrieval system, without permission in writing from the copright holder.

ABCDE
FGHIJ
KLMNO
PQRS

Printed in England

HAYNES PUBLISHING GROUP
SPARKFORD YEOVIL SOMERSET ENGLAND
distributed in the USA by
HAYNES PUBLICATIONS INC
861 LAWRENCE DRIVE
NEWBURY PARK
CALIFORNIA 91320
USA

4179218

BEDFORDSHIRE COUNTY LIBRARY COUNTY COUNCIL

629.2275 YAM

Acknowledgements

Our grateful thanks are due to R. S. Damerell and Son Ltd of St. Austell, who provided the service information necessary in the compilation of the manual, and to Fran Ridewood and Co, of Wells, who supplied the Yamaha YB100 model featured in the photographs. Jim Patch of Yeovil Motorcycle Services provided the machine featured in this manual. We are grateful also to Brian Hamilton-Farey, Service Manager of Mitsui Machinery Sales (UK) Ltd for checking the manuscript and for supplying the colour transparency used for the front cover illustration.

Brian Horsfall assisted with the stripdown and rebuilding, and devised the ingenious methods for overcoming the lack of service tools.

Les Brazier took the photographs which accompany the text. Jeff Clew edited the text.

Finally, we would like to thank the Avon Rubber Company, who kindly supplied information and technical assistance on tyre fitting; NGK Spark Plugs (UK) Ltd for information on spark plug maintenance and electrode conditions and Renold Limited for advice on chain care and renewal.

About this manual

The purpose of this manual is to present the owner with a concise and graphic guide which will enable him to tackle any operation from basic routine maintenance to a major overhaul. It has been assumed that any work would be undertaken without the luxury of a well-equipped workshop and a range of manufacturer's service tools.

To this end, the machine featured in the manual was stripped and rebuilt in out own workshop, by a team comprising a mechanic, a photographer and the author. The resulting photographic sequence depicts events as they took place, the hands shown being those of the author and the mechanic.

The use of specialised, and expensive, service tools was avoided unless their use was considered to be essential due to risk of breakage or injury. There is usually some way of improvising a method of removing a stubborn component, provided that a suitable degree of care is exercised.

The author learnt his motorcycle mechanics over a number of years, faced with the same difficulties and using similar facilities to those encountered by most owners. It is hoped that this practical experience can be passed on through the pages of this manual.

Where possible, a well-used example of the machine is chosen for the workshop project, as this highlights any areas which might be particularly prone to giving rise to problems. In this way, any such difficulties are encountered and resolved before the text is written, and the techniques used to deal with them can be incorporated in the relevant section. Armed with a working knowledge of the machine, the author undertakes a considerable amount of research in order that the maximum amount of data can be included in the manual.

Each Chapter is divided into numbered sections. Within these sections are numbered paragraphs. Cross reference throughout the manual is quite straightforward and logical. When reference is made 'See Section 6.10' it means Section 6, paragraph 10 in the same Chapter. If another Chapter were intended, the reference would read, for example. 'See Chapter 2, Section 6.10'. All the photographs are captioned with a section/paragraph number to which they refer and are relevant to the Chapter text adjacent.

Figures (usually line illustrations) appear in a logical but numerical order, within a given Chapter. Fig. 1.1 therefore refers to the first figure in Chapter 1.

Left-hand and right-hand descriptions of the machines and their components refer to the left and right of a given machine when the rider is seated normally.

Whilst every care is taken to ensure that the information in this manual is correct no liability can be accepted by the author or publishers for loss, damage or injury caused by any errors in or omissions from the information given.

Contents

Chapter	Section	Page	Section	Page
Introductory sections	Acknowledgements	2	Routine maintenance	8–13
	About this manual	2	Quick glance maintenance	
	Introduction to the Yamaha		adjustments and capacities	14
	YB100	6	Recommended torque settings	14
	Dimensions and weight	6	Recommended lubricants	14
	Ordering spare parts	7	Working conditions and tools	15
Chapter 1 Engine, clutch and gearbox	Operations with unit in frame	17	Examination and renovations:	
	Engine/gearbox removal	18	gearbox	34
	Dismantling general	21	clutch	36
	Disc valve		Unit reassembly – general	37
	Examination and renovation:		Replacing unit in frame	50
	general	30	Starting and running the	
	Decarbonising	33	rebuilt engine	53
Chapter 2 Fuel system and lubrication	Petrol tank removal	56	Air filter	62
	Fuel tap	57	Engine lubrication system	62
	Carburettor	57-62	Oil pump	64
	Disc valve	62	Silencer – decarbonising	64
Chapter 3 Ignition system	Contact breaker	68	Ignition coil	71
	Condenser	70	Sparking plug	72
Chapter 4 Frame and forks	Front forks	74,76,78	Centre stand	81
	Steering head lock	78	Prop stand	81
	Frame examination	79	Foot rests	81
	Swinging arm rear fork	79	Rear brake pedal	82
	Suspension units	80	Cleaning the machine	83
Chapter 5 Wheels, brakes and tyres	Front wheel	84	Adjusting brakes	91
	Speedometer drive gearbox	85	Cush drive assembly	93
	Rear wheel end brake	86	Rear wheel sprocket	93
	Brakes	88	Final drive chain	93
	Wheel bearings	89	Tyres	94
	Wheel bearings	89		
Chapter 6 Electrical system	Charging system – checking	98	Switches	102
	Rectifier – testing	98	Stop and tail lamp	103
	Battery	99	Flasher indication lamps	103
	Fuse locations	99	Speedometer and warning lamps	103
	Headlamp	99	Horn	106

Note: General descriptions and specifications are given in each Chapter immediately after the list of contents. Fault diagnosis is given at the end of each Chapter.

Wiring diagrams – Colour 110-111
Metric conversion tables 112-113
Index 114-116

1978 Yamaha YB100 model

Close-up of YB100 engine unit

1979 Yamaha YB100E model

Close-up of YB100E engine unit

Introduction to the Yamaha YB 100

Although the history of Yamaha can be traced back to the year 1887, when a then very small company commenced manufacture of reed organs, it was not until 1954 that the company became interested in motor cycles. As can be imagined, the problems of marketing a motor cycle against a background of musical instrument manufacture were considerable. Some local racing successes and the use of hitherto unknown bright colour schemes helped achieve the desired results and in July 1955 the Yamaha Motor Company was established as a separate entity, employing a work force of less than 100 and turning out some 300 machines a month.

Competition successes continued and with the advent of tasteful styling that followed Italian trends, Yamaha became established as one of the world's leading motor cycle manufacturers. Part of this success story is the impressive list of Yamaha 'firsts' – a whole string of innovations that include electric starting, pressed steel frames, torque induction and 6 and 8 port engines. There is also the "Autolube" system of lubrication, in which the engine-driven oil pump is linked to the twist grip throttle, so that lubrication requirements are always in step with engine demands.

Since 1964, Yamaha has gained the World Championship on numerous occasions, in both the 125 cc and 250 cc classes. Indeed, Yamaha has dominated the lightweight classes in international road racing events to such an extent in recent years that several race promoters are now instituting a special type of event in their programme from which Yamaha machines are barred! Most of the racing successes have been achieved with twin cylinder two-strokes and the practical experience gained has been applied to the road going versions.

The YB 100 models are designed to fulfil the role of an economical and efficient lightweight commuter machine. The capacity of 100 cc gives adequate performance for use in normal traffic, but without the high fuel consumption of more highly tuned two-stroke machines. Engine efficiency is enhanced by the use of a disc valve induction system, which gives more precise control of the incoming mixture.

A number of features of these machines are designed to minimise the effects of wear and tear, such as the shrouded front forks and rear suspension units, and the full chain enclosure.

Dimensions and weight

Overall length	1850 mm (72.8 in)
Overall height	1035 mm (40.7 in)
Overall width	735 mm (28.9 in)
Wheelbase	1180 mm (46.5 in)
Seat height	785 mm (30.9 in)
Dry weight	84 kg (185 lb)

Ordering spare parts

When ordering spare parts for any Yamaha, it is advisable to deal direct with an official Yamaha agent who should be able to supply most of the parts ex stock. Parts cannot be obtained from Yamaha direct and all orders must be routed via an approved agent even if the parts required are not held in stock. Always quote the engine and frame numbers in full, especially if parts are required for earlier models.

The frame and engine numbers are stamped on a Manufacturer's Plate riveted to the steering head on the left-hand side. The frame number is also stamped on the frame itself on the right-hand side of the steering head. The engine number is stamped on the upper crankcase.

Use only genuine Yamaha spares. Some pattern parts are available that are made in Japan and may be packed in similar looking packages. They should be only be used if genuine parts are hard to obtain or in an emergency, for they do not normally last as long as genuine parts, even though there may be a price advantage.

Some of the more expendable parts such as sparking plugs, bulbs, tyres, oils and greases etc., can be obtained from accessory shops and motor factors, who have convenient opening hours, and can often be found not far from home. It is also possible to obtain parts on a Mail Order basis from a number of specialists who advertise regularly in the motorcycle magazines.

Location of Frame No

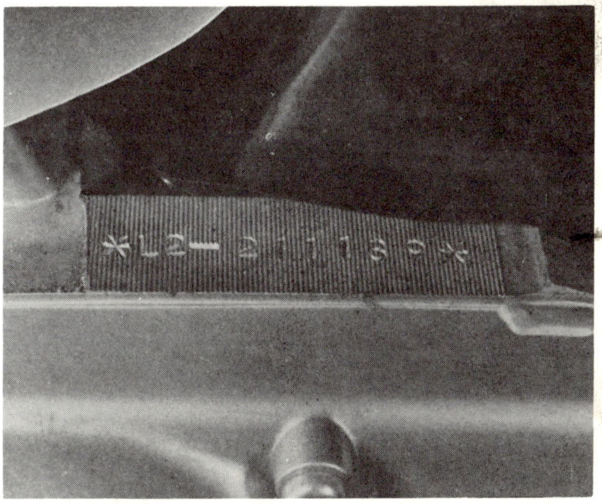

Location of Engine No

Routine maintenance

Introduction

Periodic routine maintenance is a continuous process that commences immediately the machine is used. It must be carried out at specified mileage readings, or on a calendar basis if the machine is not used frequently, whichever is the sooner. Maintenance should be regarded as an insurance policy, to help keep the machine in the peak of condition and to ensure long, trouble-free service. It has the additional benefit of giving early warning of any faults that may develop and will act as a regular safety check, to the obvious advantage of both rider and machine alike.

The various maintenance tasks are described under their respective mileage and calendar headings. Accompanying diagrams are provided, where necessary. It should be remembered that the interval between the various maintenance tasks serves only as a guide. As the machine gets older or is used under particularly adverse conditions, it would be advisable to reduce the period between each check.

For ease of reference each service operation is described in detail under the relevant heading. However, if further general information is required, it can be found within the manual under the pertinent section heading in the relevant Chapter.

In order that the routine maintenance tasks are carried out with as much ease as possible, it is essential that a good selection of general workshop tools are available.

Included in the kit must be a range of metric ring or combination spanners, a selection of crosshead screwdrivers and at least one pair of circlip pliers.

Additionally, owing to the extreme tightness of most casing screws on Japanese machines, an impact screwdriver, together with a choice of large or small crosshead screw bits, is absolutely indispensable. This is particularly so if the engine has not been dismantled since leaving the factory.

Weekly or every 200 miles (300 km)

1 *Topping up engine oil*

The oil tank level should be checked on a daily basis, prior to starting the engine. A small plastic sight glass or level gauge gives an immediate visual warning of whether the oil level has dropped too low. Although it is quite safe to use the machine as long as oil is visible in the sight glass, it is recommended that the level is maintained to within about an inch of the tank filler neck, to allow a good reserve. It is advised that the tank is topped up at weekly intervals.

Early models have sight glass in oil tank, ...

... whilst later machines have level gauge

Tank should be filled to within 1 inch of neck

Electrolyte level should be between lines on case

2 *Tyre pressures*

It is essential that the tyres are kept inflated to the correct pressure at all times. Under or over-inflated tyres can lead to accelerated rates of wear, and more importantly, can render the machine inherently unsafe. Whilst this may not be obvious during normal riding, it can become painfully and expensively so in an emergency situation, as the tyres' adhesion limits will be greatly reduced.

Check the tyre pressures with a pressure gauge that is known to be accurate. Always check the pressure when the tyres are cold. If the machine has travelled a number of miles, the tyres will have become hot and consequently the pressure will have increased. A false reading will therefore result.

It is recommended that a small pocket gauge is purchased and carried on the machine, as the readings on garage forecourt gauges can vary and may often be inaccurate.

The pressures given are those recommended for the tyres fitted as original equipment. If replacement tyres are purchased, the pressure settings may vary. Any reputable tyre distributor will be able to give this information.

Tyre pressures	*Solo*	*With pillion passenger*
Front	21 psi (1.5 kg cm ^{2})	26 psi (1.8 kg cm^2)
Rear	28 psi (2.0 kg cm^2)	33 psi (2.3 kg cm^2)

3 *Battery electrolyte level*

A GS battery is fitted as standard equipment to the YB 100 models. The battery is of the conventional lead-acid type and has a capacity of 4 ampere hours.

The transparent plastic case of the battery permits the upper and lower levels of the electrolyte to be observed when the left-hand side panel has been removed. Maintenance is normally limited to keeping the electrolyte level between the prescribed upper and lower limits and by making sure that the vent pipe is not blocked. The lead plates and their separators can be seen through the transparent case, a further guide to the general condition of the battery.

Unless acid is spilt, as may occur if the machine falls over, the electrolyte should always be topped up with distilled water, to restore the correct level. If acid is spilt on any part of the machine it should be neutralised with an alkali such as washing soda and washed away with plenty of water, otherwise serious corrosion will occur. Top up with sulphuric acid of the correct specific gravity (1.260 – 1.280) only when spillage has occurred. Check that the vent pipe is well clear of the frame tubes or any of the other cycle parts, for obvious reasons.

4 *Control cable lubrication*

Apply a few drops of motor oil to the exposed inner portion of each control cable. This will prevent drying-up of the cables between the more thorough lubrication that should be carried out during the 2000 mile/3 monthly service.

5 *Rear chain lubrication and adjustment*

In order that the life of the rear chain be extended as much as possible, regular lubrication and adjustment is essential.

Intermediate lubrication should take place at the weekly or 200 mile service interval with the chain in situ. Application of one of the proprietary chain greases contained in an aerosol can is ideal. Ordinary engine oil can be used, though owing to the speed with which it is flung off the rotating chain, its effectiveness is limited.

The chain lubricant may be applied via the inspection hole in the chain enclosure, where this is fitted.

Adjust the chain after lubrication, so that there is approximately 20 mm ($\frac{3}{4}$ in) slack in the middle of the lower run. Always check with the chain at the tightest point as a chain rarely wears evenly during service.

Adjustment is accomplished after placing the machine on the centre stand and slackening the wheel nuts, so that the wheel can be drawn backwards by means of the drawbolt adjusters in the fork ends.

The torque arm nuts and the rear brake adjuster must also be slackened during this operation. Adjust the drawbolts an equal amount to preserve wheel alignment. The fork ends are clearly marked with a series of parallel lines above the adjusters, to provide a simple visual check.

6 *Safety check*

Give the machine a close visual inspection, checking for loose nuts and fittings, frayed control cables etc. Check the tyres for damage, especially splitting on sidewalls. Remove any stones or other objects caught between the treads. This is particularly important on the front tyre, where rapid deflation due to penetration of the inner tube will almost certainly cause total loss of control.

7 *Legal check*

Ensure that the lights, horn and trafficators function correctly, also the speedometer.

Use aerosol chain grease between full lubrication intervals

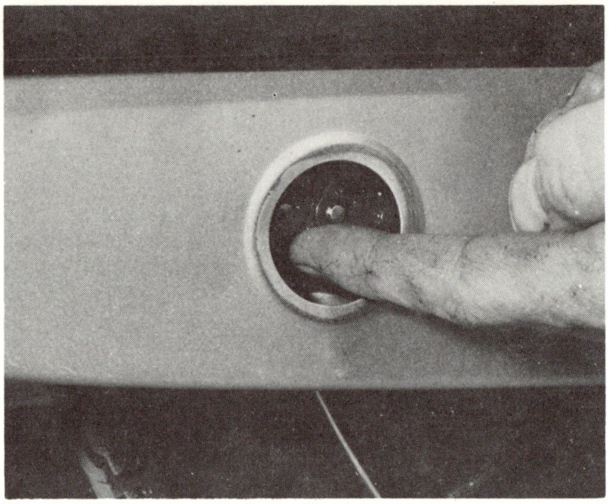
Check chain free play via inspection hole

Marks on swinging arm aid wheel alignment

Three monthly or every 2000 miles (3000 km)

Complete all the checks listed in the weekly/200 mile service and then the following items:

1 *Oil pump adjustment*

The oil pump adjustment should be checked every three months to ensure that the correct amount of oil is being delivered to the engine. The pump setting can be checked after the right-hand outer casing has been removed. Start by setting the throttle cable free play, then check that the mark on the pump pulley aligns with the pin on the pump boss, when the throttle is opened to the correct position. A small circular mark on the throttle valve should just touch the top of the carburettor bore at this setting. See Chapter 2, Section 11, for full details.

2 *Checking the sparking plug*

Pull off the sparking plug cap and using the correct size plug spanner, remove the plug. Clean off any carbon or oil from the electrodes and using a feeler gauge, check the gap. Reset the gap, if necessary, after referring to Chapter 3. Refit the sparking plug into the cylinder head but do not overtighten it, as stripping of the threads could result. Refit the plug cap.

A new sparking plug should be fitted every 6000 miles (9000 km), or earlier if it is excessively worn, burnt or damaged.

The sparking plug gap should be set to 0.6 – 0.7 mm (0.024 – 0.028 in).

3 *Adjust slow running speed*

Adjust the slow running speed only if necessary. Refer to Chapter 2, Section 8.

4 *Check and adjust the contact breaker gap*

The correct setting of the contact breaker gap is critical and engine performance can be adversely affected if it is not carried out correctly. Refer to Chapter 3, Section 2 for the relevant details.

5 *Oil the contact breaker cam felt pad*

When checking the contact breaker points gap, the felt pad can be seen through one of the flywheel apertures. Lubricate the felt pad with a few drops of light machine oil to reduce wear on the heel of the points. Do not over oil, or excess oil will find its way on to the points and cause ignition problems.

6 *Clean the fuel tap filter*

Ensure the fuel tap filter is clean so that a smooth flow of fuel passes through the fuel tap. Turn the fuel tap to the OFF position. Using a 10 mm spanner, remove the cup and O-ring. Remove the filter gauze and wash it in clean petrol. Refit the filter O-ring and cup. Tighten the cup, using a 10 mm spanner.

7 *Exhaust pipe ring nuts*

Check the tightness of the ring nuts at the cylinder barrel and at the silencer joint.

8 *Clutch adjustment*

Check the clutch cable free play and release mechanism adjustment as described below:

Slacken the adjuster locknut by about one turn, and hold this position with a spanner. Slowly screw the adjuster inwards until some resistance is felt, indicating that all free play has been taken up. If necessary, experiment a little to get the feel of this point. The screw should be backed off $\frac{1}{4}$ turn from the above point, and then held in this position while the locknut is retightened. Moving to the cable adjuster on the outside of the casing, slacken the locknut and set the adjuster to give 2 – 3 mm (0.08 – 0.12 in) free play measured between the handlebar lever stock and blade.

Slacken locknut and set adjusting screw ...

... then adjust cable free play here

9 *Clean the air filter*

If the air cleaner filter becomes blocked, intake resistance increases, resulting in loss of power and in increase in fuel consumption.

Release the domed nut or screw which secures the chromium-plated cover to the left-hand end of the air filter casing. Lift the cover away, and withdraw the filter element. The element should be washed in a degreasing solvent or in petrol to remove all the accumulated dust and the old oil. Squeeze out the solvent and allow the element to dry, then re-impregnate it with clean engine oil. The filter should be soaked evenly in oil, but not to the extent that it is dripping with it.

Note that the air filter element must be renewed if it has become torn or damaged in any way. Check that the intake hoses are not split or perished, renewing where necessary.

10 *Change the gearbox/primary drive oil*

It is better to change the transmission oil when it is hot, and the viscosity is low.

Place the machine on its centre stand and position a suitable container under the gearbox oil drain plug. Remove the plug and the oil filler cap and allow the oil to drain. Check the condition of the drain plug sealing washer and when the flow of oil has ceased, wipe the area around the hole, refit and tighten the plug.

Fill the gearbox with sufficient SAE 10W/30 engine oil to reach midway between the upper and lower level marks on the dipstick. Note that the measurement is made with the filler cap resting in position, and *not* screwed fully home. The casing will accept about 600 cc (1.06 Imp pint, 1.48 US pint) at oil changes.

11 *General lubrication*

Apply grease or oil to the handlebar lever pivots and to the centre stand and prop stand pivots.

12 *Control cable lubrication*

Lubricate the control cables thoroughly with motor oil or an all-purpose oil. A good method of lubricating the cables is shown in the accompanying illustration, using a plasticine funnel. This method has the disadvantage that the cables usually need removing from the machine. An hydraulic cable oiler which pressurises the lubricant overcomes this problem. Do not lubricate nylon lined cables (which may have been fitted as replacements), as the oil may cause the nylon to swell, thereby causing total cable seizure.

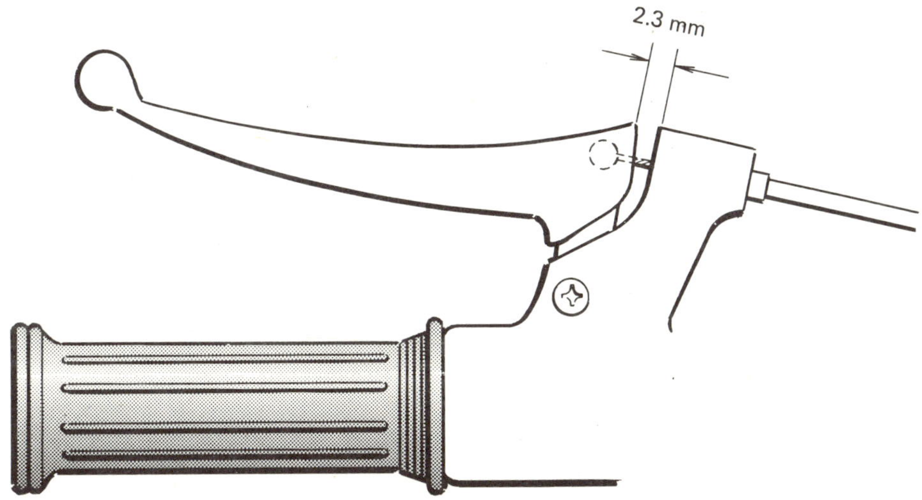

Clutch cable adjustment

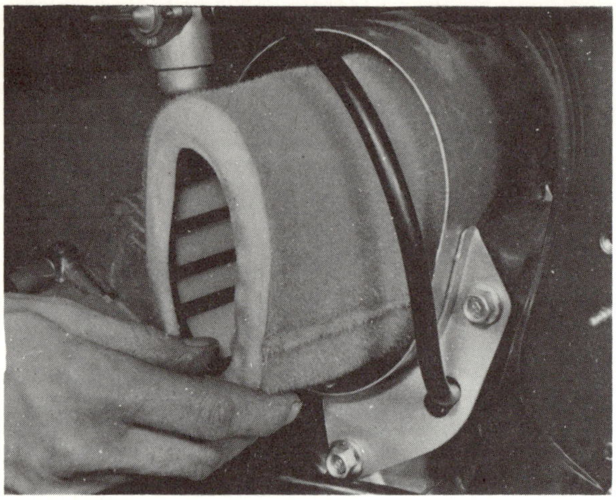

Element can be removed for cleaning

Top up the transmission oil level ...

... to within upper and lower level marks

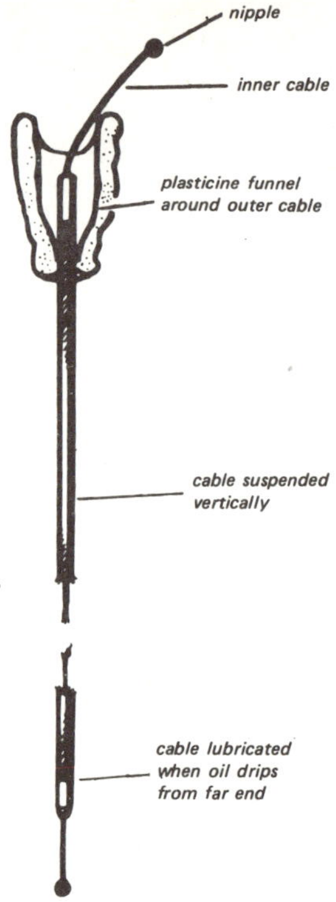

Control cable lubrication

13 Removing and relubricating the final drive chain

The final drive chain on YB 100 models enjoys an unusually long and comfortable life, by virtue of the full enclosure.

It is still important that the chain be removed for a thorough cleaning and relubrication at six monthly intervals.

Remove the chain enclosure halves and detach the left-hand casing to gain access to the gearbox sprocket. Separate the chain by prising off the spring link and sliding the chain ends apart. The chain can then be run off the sprockets.

Wash the chain carefully in paraffin (kerosene) using a stiff brush to remove all traces of road dirt. The chain should now be rinsed in petrol (gasoline) and hung up to dry off, or blown dry with compressed air.

The cleaned chain should be checked for wear by measuring the amount of stretch which has taken place. Lay the chain lengthways in a straight line and compress it at each end to take up all play. Anchor one end and pull on the other end to extend the chain and take up all play in the other direction. If the chain extends by more than the distance between two of the rollers it should be renewed and a close examination of both the engine and rear wheel sprockets must be carried out, to check for wear and sprocket tooth damage.

The chain must be lubricated after cleaning, by immersing it in a molten chain lubricant, such as Linklyfe or Chainguard and then hanging it up to drain. This will ensure good penetration of the lubricant between the pins and rollers, making it less likely to be thrown off when the chain is in motion.

Position the two ends of the chain on the rear wheel sprocket, insert the link, fit the side plate and secure with the spring clip. Note that the closed end of the spring clip must be fitted pointing in the direction of motion.

Adjust the chain tension as previously stated in the weekly/200 mile service.

Closed end of clip must face in direction of chain travel

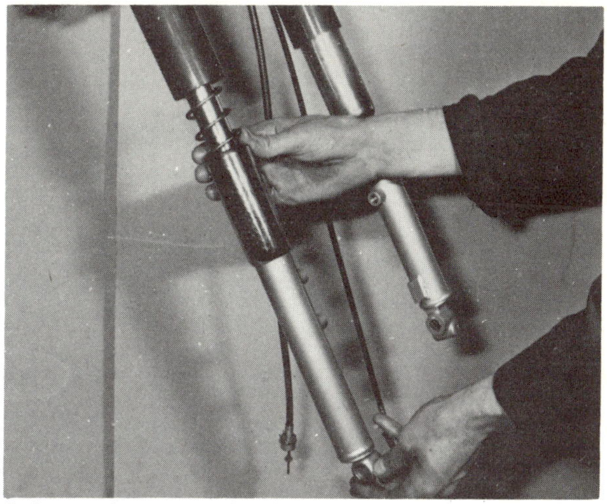

Forks must be removed for oil change

Six monthly or every 4000 miles (6000 km)

Complete the checks listed under the weekly/200 mile and three monthly/2000 mile headings, then complete the following additional procedures.

1 Changing the front fork damping oil

Over a period of time, the fork damping oil will deteriorate due to contamination by water and ageing. To offset any loss of damping effect, the oil should be changed every 4000 miles as a precautionary measure. Unfortunately, no fork drain plugs are provided, and to drain the old oil it will be necessary to remove the fork legs and empty the contents from the top of the stanchion.

As this is a fairly lengthy process, reference should be made to Chapter 4 Section 3 for a full description of the procedure involved. Once the legs have been removed, invert them over a bowl or drain tray, 'pumping' them to expel the old oil Refill each leg with 165 cc of SAE 15 engine oil or good quality fork oil.

2 Check and re-grease the steering head bearings

While the forks are removed for the oil to be changed, dismantle the steering head assembly as described in Chapter 4, Section 4. Clean the steering head bearings and examine them for wear or damage, renewing parts as necessary. If the bearings are in good condition, reassemble the unit, lubricating the bearing races with a general purpose grease. Do not omit to adjust the steering head assembly before the machine is used.

3 Lubricate the swinging arm pivot bolt

Slacken the nut on the end of the swinging arm pivot, then carefully withdraw the latter, using a screwdriver or long bolt as a means of temporarily locating the swinging arm assembly. The pivot bolt should be coated with grease to inhibit corrosion, and then refitted. Tighten the pivot bolt to 4.5 kgf m (32 lbf ft).

4 Decarbonising the cylinder head, barrel and piston

Removal of the cylinder components, decarbonising and inspection, should be carried out by referring to the relevant Sections in Chapter 1. This work can be accomplished without removing the engine from the frame.

5 Removal, inspection and relubrication of wheel bearings

Carry out the operations listed in the heading by following the procedures given in Chapter 5, Section 7.

6 Check the condition of the brake linings

Although a brake lining wear indicator is provided on some models, it is advisable to remove both wheels from the machine, in turn, in order to check the condition of the brake drum and the extent to which the brake linings have worn. Since this will entail a small amount of dismantling, reference should be made to Chapter 5 for information about the removal of the front wheel (Section 3) or the rear wheel (Section 5). Visual inspection will show whether the brake linings require attention, and it is recommended that reference is made to Section 6 of the same Chapter if it is necessary to remove the brake shoe assembly from either wheel.

Twelve monthly or every 8 000 miles (12 000 km)

Complete all the checks listed under the weekly, monthly, three and six monthly headings, but only if they are not directly connected with the tasks listed below. More extensive dismantling is required when undertaking these latter tasks and reference to the relevant Chapters and Sections will be necessary in each case:

1 Dismantle, clean, examine and reassemble the carburettor.
2 Renew the contact breaker assembly.
3 Decarbonise the engine and clean the exhaust system, especially the silencer.

Quick glance maintenance adjustments and capacities

Engine (oil tank)	2-stroke engine oil. Fill to within 1 in of filler neck	
Gearbox	SAE 10W/30 engine oil – 600 cc at oil change, 700 cc dry	
Contact breaker gap	0.3–0.4 mm (0.012–0.016 in)	
Sparking plug gap	0.6–0.7 mm (0.024–0.028 in)	
Tyre pressures:	Solo	Pillion
Front	21 psi (1.5 kg cm^2)	26 psi (1.8 kg cm^2)
Rear	28 psi (2.0 kg cm^2)	33 psi (2.3 kg cm^2)

Recommended torque settings

Cylinder head nuts	1.4 kgf m (10 lbf ft)
Generator flywheel nuts	5.0 kgf m (36 lbf ft)
Gearbox sprocket nut	5.5 kgf m (40 lbf ft)
Crankcase screws	0.8 kgf m (6 lbf ft)
Crankshaft pinion nut	5.9 kgf m (43 lbf ft)
Front wheel spindle nut	5.9 kgf m (43 lbf ft)
Swinging arm pivot shaft	4.5 kgf m (32 lbf ft)
Rear wheel spindle nut	5.9 kgf m (43 lbf ft)

Recommended lubricants

Component	Type and specification
Engine	Yamaha Autolube oil or equivalent two-stroke engine oil
Gearbox	SAE 10W/30 engine oil
Contact breaker pivot pin	One or two drops of light machine oil
Contact breaker cam lubricating wick	One or two drops of light machine oil
Front forks	165 cc Fork oil or SAE15 engine oil per leg
Steering head bearings	Medium weight grease, Castrol LM or similar
Wheel bearings	High melting point grease
Brake pivots	High melting point grease
Electrical contacts	WD 40 aerosol, or equivalent
Chain	Aerosol chain lubricant, or hot immersion lubricant such as Linklyfe or Chainguard

Working conditions and tools

When a major overhaul is contemplated, it is important that a clean, well-lit working surface is available, equipped with a workbench and vice, and with space for laying out or storing the dismantled assemblies in an orderly manner where they are unlikely to be disturbed. The use of a good workshop will give the satisfaction of work done in comfort and without haste, where there is little chance of the machine being dismantled and reassembled in anything other than clean surroundings.Unfortunately, these ideal working conditions are not always practicable and under these latter circumstances when improvisation is called for, extra care and time will be needed.

The other essential requirement is a comprehensive set of good quality tools. Quality is of the prime importance since cheap tools will prove expensive in the long run if they slip or break and damage the components to which they are applied. A good quality tool will last a long time, and more than justify the cost. The basis of any tool kit is a set of open-ended spanners, which can be used on almost any part of the machine to which there is reasonable access. A set of ring spanners makes a useful addition, since they can be used on nuts that are very tight or where access is restricted. Where the cost has to be kept within reasonable bounds, a compromise can be effected with a set of combination spanners – open-ended at one end and having a ring of the same size on the other end. Socket spanners may also be considered a good investment, a basic $\frac{3}{8}$ in or $\frac{1}{2}$ in drive kit comprising a ratchet handle and a small number of socket heads, if money is limited. Additional sockets can be purchased, as and when they are required. Provided they are slim in profile, sockets will reach nuts or bolts that are deeply recessed. When purchasing spanners of any kind, make sure the correct size standard is purchased. Almost all machines manufactured outside the UK and the USA have metric nuts and bolts. whilst those produced in Britain have BSF or BSW sizes. The standard used in the USA is AF, which is also found on some of the later British machines. Other tools that should be included in the kit are a range of crosshead screwdrivers, a pair of pliers and a hammer.

When considering the purchase of tools, it should be remembered that by carrying out the work oneself, a larger proportion of the normal repair cost, made up by labour charges, will be saved. The economy made on even a minor overhaul will go a long way towards the improvement of a tool kit.

In addition to the basic tool kit, certain additional tools can prove invaluable when they are close to hand, to help you speed up a multitude of repetitive jobs. For example, an impact screwdriver will ease the removal of screws that have been tightened by a similar tool, during assembly, without a risk of damaging the screw heads. And, of course, it can be used again to retighten the screws, to ensure an oil or airtight seal results. Circlip pliers have their uses too, since gear pinions, shafts and similar components are frequently retained by circlips that are not too easily displaced by a screwdriver. There are two types of circlip pliers, one for internal and one for external circlips. They may also have straight or right-angled jaws.

One of the most useful of all tools is the torque wrench, a form of spanner that can be adjusted to slip when a measured amount of force is applied to any bolt or nut. Torque wrench settings are given in almost every modern workshop or service manual, where the extent is given to which a complex component, such as a cylinder head, can be tightened without fear of distortion or leakage. The tightening of bearing caps is yet another example. Overtightening will stretch or even break bolts, necessitating extra work to extract the broken portions.

As may be expected the more sophisticated the machine, the greater is the number of tools likely to be required if it is to be kept in first class condition by the home mechanic. Unfortunately, there are certain jobs which cannot be accomplished successfully without the correct equipment and although there is invariably a specialist who will undertake the work for a fee, the home mechanic will have to dig more deeply in his pocket for the purchase of similar equipment if he does not wish to employ the services of others. Here a word of caution is necessary, since some of these jobs are best left to the expert. Although an electrical multimeter of the AVO type will prove helpful in tracing electrical faults, in inexperienced hands it may irrevocably damage some of the electrical components if a test current is passed through them in the wrong direction. This can apply to the synchronisation of twin or multiple carburettors too, where a certain amount of expertise is needed when setting them up with vacuum gauges. These are, however, exceptions. Some instruments such as a strobe lamp are virtually essential when checking the timing of a machine powered by CDI ignition system. In short, do not purchase any of these special items unless you have the experience to use them correctly.

Although this manual shows how components can be removed and replaced without the use of special service tools (unless absolutely essential), it is worthwhile giving consideration to the purchase of the more commonly used tools if the machine is regarded as a long term purchase. Whilst the alternative methods suggested will remove and replace parts without risk of damage, the use of the special tools recommended and sold by the manufacturer will invariably save time.

Chapter 1 Engine, clutch and gearbox

Contents

General description	1
Operations with the engine/gearbox in the frame	2
Engine and gearbox unit: removal from the frame	3
Dismantling the engine, clutch and gearbox: general	4
Preventing the engine from turning, both for dismantling and reassembly purposes	5
Cylinder head: removal	6
Cylinder barrel: removal	7
Piston and small end: removal	8
Oil pump: removal	9
Gearbox sprocket: removal	10
Right-hand inner cover: removal	11
Clutch: removal	12
Kickstart mechanism: removal	13
Gear selector mechanism: removal	14
Disc valve assembly: removal	15
Flywheel generator assembly: removal	16
Separating the crankcase halves	17
Gearbox components and crankshaft: removal	18
Examination and renovation: general	19
Engine casings, bearings and oil seals: examination and renovation	20
Crankshaft assembly: examination and renovation	21
Decarbonising	22
Cylinder head: examination and renovation	23
Cylinder barrel and bore: examination and renovation	24
Piston and piston rings: examination and renovation	25
Piston/bore clearance check	26
Gearbox components: examination and renovation	27
Clutch assembly: examination and renovation	28
Gearbox components: reassembling the gear clusters	29
Engine and gearbox reassembly: general	30
Engine and gearbox reassembly: crankcase preparation	31
Engine and gearbox reassembly: joining the crankcase halves	32
Engine and gearbox reassembly: fitting the disc valve	33
Engine and gearbox reassembly: refitting the selector mechanism	34
Engine and gearbox reassembly: refitting the kickstart mechanism	35
Engine and gearbox reassembly: refitting the clutch and crankshaft pinion	36
Engine and gearbox reassembly: refitting the right-hand inner cover	37
Engine and gearbox reassembly: refitting the flywheel generator	38
Piston, cylinder barrel and cylinder head: reassembly	39
Replacing the engine unit in the frame	40
Starting and running the rebuilt engine	41
Fault diagnosis: engine	42
Fault diagnosis: clutch	43
Fault diagnosis: gearbox	44

Specifications

Engine

Type	Single cylinder, air-cooled two-stroke
Induction timing	By crankshaft-mounted disc valve
Capacity	97 cc (5.92 cu in)
Bore	52.0 mm (2.05 in)
Stroke	45.6 mm (1.79 in)
Compression ratio	6.5:1
Lubrication	By pump (Yamaha Autolube)

Piston

Type	Light alloy
Clearance in bore:	
Minimum	0.030 mm (0.0012 in)
Maximum	0.035 mm (0.0014 in)
Oversizes available:	
1st oversize	52.25 mm (2.057 in)
2nd oversize	52.50 mm (2.067 in)
3rd oversize	52.75 mm (2.077 in)
4th oversize	53.00 mm (2.086 in)

Piston rings	
Type	Keystone
End gap:	
Minimum	0.15 mm (0.0059 in)
Maximum	0.35 mm (0.0138 in)
Small end bearing	
Type	Needle roller
Gudgeon pin diameter	14 mm (0.55 in)
Big end bearing	
Type	Needle roller
Crankshaft assembly	
Mainshaft runout	0.03 mm (0.0012 in) maximum
Connecting rod axial clearance	0.2–0.7 mm (0.008–0.027 in)
Maximum axial movement at small end eye	0.8–2.0 mm (0.032–0.079 in)
Crankpin outside diameter	22 mm (0.87 in)
Crankpin length	48.7 mm (1.92 in)
Clutch	
Type	Wet, multiplate
No of plain plates	3
No of friction plates	4
Friction plate thickness	3.5 mm (0.138 in) nominal
Friction plate wear limit	3.2 mm (0.126 in)
Plain plate maximum warpage	0.05 mm (0.002 in)
Springs	6
Free length	28.2 mm (1.11 in)
Wear limit	27.2 mm (1.07 in)
Release mechanism	Quick screw type
Gearbox	
Type	4-speed constant mesh
Ratios:	
1st gear	3.077:1 (40T:13T)
2nd gear	1.889:1 (34T:18T)
3rd gear	1.304:1 (30T:23T)
Top gear	0.963:1 (26T:27T)
Kickstart mechanism	Primary type
Primary drive	
Type	Helical gear
Ratio	3.895:1 (74T:19T)
Secondary drive	
Type	Chain
Ratio	2.4:1 (36T:15T)

1 General description

The Yamaha YB100 is equipped with a single cylinder air-cooled two-stroke engine unit suspended from its spine frame. The cylinder barrel is inclined forwards keeping the engine weight, and consequently, the machine's centre of gravity, low. The engine drives through a conventional multiplate clutch and gear primary transmission to a foot operated four-speed gearbox built in unit with the engine.

The engine is lubricated by a pump delivery system known as Yamaha Autolube, in which oil is fed to the engine on a constant-loss basis by an engine driven oil pump, from a frame mounted oil tank, the delivery rate being matched to the throttle opening.

The clutch and gearbox components are remote from the crankcase proper, being housed in integral casings formed by extensions of the crankcase castings. The various components are lubricated by an oil bath arrangement.

Induction is controlled by a rotary valve arrangement in the interests of economy and improved torque characteristics. The valve comprises a fibre or nylon disc which is attached to the right-hand crankshaft end and which is enclosed in a close-fitting housing. A cut-out in the disc aligns with the intake port at a predetermined point, allowing the intake of the incoming mixture to be closely controlled.

Note: Before any overhaul or servicing work is undertaken, it should be noted that one of the manufacturer's Service Tools should be considered essential, and is well worth buying if the machine is to be kept for any length of time. The tool is an extractor for the generator rotor, and is dealt with in detail in Section 16 of this Chapter.

2 Operations with the engine/gearbox in the frame

It is not necessary to remove the engine/gearbox unit from the frame unless the crankshaft assembly and/or gearbox components need attention. Most operations can be accomplished with the engine/gearbox in place, such as:

a) Removal and installation of the cylinder head.
b) Removal and installation of the cylinder barrel and piston.
c) Removal and installation of the flywheel generator and contact breaker assembly.
d) Removal and installation of the clutch.
e) Removal and installation of the kickstart spring.

When several operations need to be undertaken at the same time, it would probably be advantageous to remove the complete unit from the frame, a comparatively simple operation. This will afford better access and more working space.

3 Engine and gearbox unit: removal from the frame

1 Place the machine on its centre stand so that it is supported firmly, and allowing adequate working clearance on each side. Make sure that the petrol tap is in the 'off' position, then prise off the petrol feed pipe, using a small screwdriver. The engine unit can be removed without disturbing the petrol tank, provided that care is taken not to damage the paint finish. If it is thought preferable to remove the tank to obviate this risk, note that a connecting tube runs between the two tank halves, and this must be prised off before the tank is removed. As a consequence of this, it will be necessary to drain the tank's contents before the connecting tube is released.

2 Release the battery and toolbox cover on the left-hand side of the machine, and disconnect and remove the battery to prevent any possible short circuit occurring during engine removal. Moving to the right-hand side of the machine, release the four screws which retain the right-hand engine casing. These tend to be tight, and if so, it is preferable to use an impact screwdriver to avoid damaging the screw heads. Lift the cover away to gain access to the carburettor and oil pump. The carburettor can be removed as an assembly and left attached to its control cables unless specific attention to the instrument is required.

3 Prise out the rubber blanking plug at the front of the casing so that a screwdriver can be introduced to slacken the carburettor clamp screw. With the screw loosened, pull the carburettor off its stub so that the petrol feed, overflow and breather tubes can be disconnected. Check that the petrol tap is turned off first, if the petrol tank is still in position. On early models, the rubber shroud or cover through which the carburettor controls pass is a sliding fit in the casing, and will be displaced as the carburettor is withdrawn. On later units, a modified cover is fitted, which is retained by a pressed steel surround and four screws. In this case, the four screws should be removed to permit the surround and the rubber cover to be slid upwards, leaving them in place on the cables. The carburettor can be withdrawn together with the rubber cover and cables, and the assembly can then be tied clear of the engine unit.

4 Slacken off the clutch cable adjuster locknut and screw the adjuster fully inwards. Slacken and remove the clutch release mechanism adjuster locknut at the end of the actuating lever. This will allow the lever to be pulled off and disengaged from the actuating lever end. The adjuster can now be unscrewed completely to free the cable from the casing. The actuating lever can be temporarily refitted to prevent its subsequent loss.

5 Slacken the oil pump control cable adjuster locknut, and unscrew the adjuster to free it from its stop. The cable end can then be freed from the oil pump pulley, and the clutch and oil pump cables tied clear of the engine unit.

6 Using the 'C' spanner supplied in the toolkit, slacken the large sleeve nut which retains the exhaust pipe to the cylinder barrel, and also the nut which secures the exhaust pipe to silencer joint. If the correct 'C' spanner is not available, a strap wrench makes a good substitute. Failing that, a large worm drive hose clip can be fitted round the nut as a means of obtaining purchase with a Mole type wrench or similar.

7 With the two sleeve nuts slackened, the exhaust pipe can be lifted away, leaving the silencer in position. Obtain a container of at least 700 cc (1.25 Imp pints) capacity into which the transmission oil can be drained. Remove the large drain plug from the underside of the engine unit, and leave the oil to drain thoroughly whilst work is carried out elsewhere. When all of the old oil has drained, clean the drain plug and orifice, and refit the plug using a new sealing washer. Tighten the plug to 2.0 kgf m (14.0 lbf ft).

8 Release the clamp bolt which retains the gearchange pedal to its splined shaft, and pull the pedal free. Release the three screws which retain the left-hand engine casing, and lift it away. Note that it is not necessary to remove the circular inspection cover. A decision must now be taken about the method of disconnecting the generator output leads. If the unit is to be removed and the crankcases separated it will be necessary to remove the generator assembly. If this is the case, it is recommended that this be done now, and the generator assembly lodged or tied clear of the engine unit. This will obviate the need, on earlier models, to trace the leads back into the frame. They can be disconnected here, but access is difficult, especially during reassembly. If it is decided that the generator is to be removed, follow the removal instructions given in Section 16, noting that engine rotation can be prevented by selecting top gear and applying the rear brake whilst the nut is slackened. If it is decided that the generator is to be left in position, further instructions will be found in paragraph 12 of this Section.

9 It is possible to disconnect the rear chain and complete removal without disturbing the rear chain enclosure. It is preferable, however, to remove the enclosure to gain better access and to facilitate reassembly. The enclosure is in two halves, each of which is retained by two cross-head screws. Note that the rearmost screw on the upper half passes through the swinging arm web and then into the chaincase. With the screws removed, release the single screw and nut which holds the two halves together. This is located at the rear of the enclosure, immediately adjacent to the wheel. Lift away the two halves of the chain enclosure to expose the final drive chain. If the gearbox sprocket is to be removed, it is a good idea to slacken the retaining nut at this stage. Knock back the tab washer which retains the nut. Slacken the sprocket retaining nut whilst holding the rear brake on to prevent the sprocket from rotating.

10 If the chain enclosure has been removed, turn the rear wheel until the joining link appears at the rear wheel sprocket, then prise off the spring clip using a pair of pointed-nosed pliers. The link can now be slid apart and the chain ends separated. Reassemble the link on one end of the chain to prevent possible loss, then run the chain off the sprockets. It is worthwhile taking this opportunity of cleaning and relubricating the chain.

11 If it is decided to leave the chain enclosure in position, disconnect the chain at the gearbox sprocket. Take care that the ends of the chain do not disappear up inside the enclosure. This can be prevented by passing some wire through the chain and anchoring it to some convenient point on the frame.

12 If the engine unit is to be removed with the generator assembly in position, it will be necessary to trace the output leads back to their connectors. These are located inside the frame on early models, and may be reached via the aperture on the left-hand side of the frame, directly behind the battery tray. On early models, a welded pressed-steel bracket is mounted in this aperture, forming part of the battery holder, and doubling as a mounting for the indicator relay. It is secured by two screws and may be removed after these have been released. The later E models make use of a modified bracket arrangement, and the generator connector plugs into this. Where the early type is used, the bracket should be released and can be left to hang on the various leads. The generator output lead connectors can be reached through the aperture and separated. Pull the leads downwards to clear the bottom of the frame member. It may prove easier to leave this operation until the engine can be pivoted downwards to provide better clearance.

13 Release the single bolt on the left-hand side of the machine which secures the footrest assembly to the frame. The lower rear engine mounting bolt also locates the footrest assembly, and this must be removed after the mounting bolt has been withdrawn. Remove the footrest assembly together with the prop stand, then temporarily refit the lower rear mounting bolt so that the engine may be pivoted downwards during removal.

14 Obtain a short length of wooden dowel with which the oil tank outlet stub may be plugged to prevent the oil content escaping. Alternatively, a short length of plastic tubing may be used if the end is folded over and secured with wire or a hose clip. Prise off the oil feed pipe and plug the outlet. Pull off the

sparking plug cap and lodge it clear of the engine unit.

15 It will be necessary to remove the complete air filter assembly from the machine. Start by releasing the connecting hose between the engine casing and the air cleaner body. On early models, the engine end of the hose is secured by a retaining plate and screws, whilst on late models, a conventional hose clip is employed. On all models, the upper end of the hose is secured by a hose clip. The air filter case, or body, is mounted by means of a bracket. One of the mounting bolts passes through the frame, the lower one being the front engine mounting bolt.

16 Remove the two upper mounting bolts and allow the engine unit to pivot downwards, arranging suitable wooden blocks to support it. With the engine unit in this position it will be possible to reach the neutral light switch at the rear of the left-hand crankcase half. Slacken the screw which retains the neutral light lead to the switch, and release the lead. Make a final check around the engine unit to ensure that nothing remains connected which would impede removal.

17 It is helpful to have some assistance at this stage, but by no means essential. Support the engine unit whilst the remaining mounting bolt is withdrawn, then lower the rear of the unit to clear the frame and lift it away. The unit is not unduly heavy and can be managed comfortably by one person. Place the unit on a workbench to await further dismantling.

3.3 Carburettor clamp screw can be reached via access hole

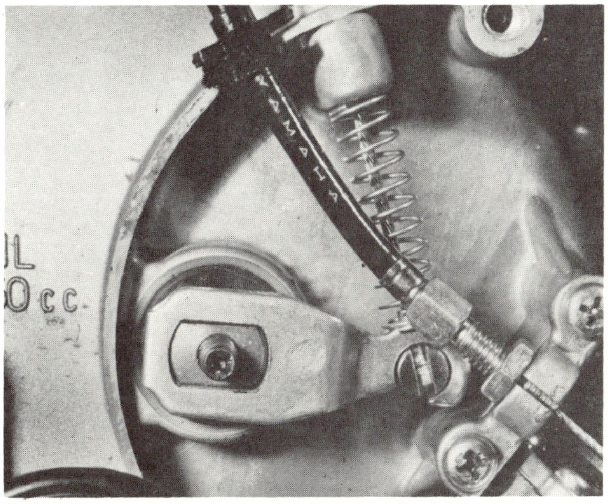

3.4 Release and detach the clutch arm and cable ...

3.5 ... followed by the oil pump control cable

3.9a Chain enclosure is secured by four screws (arrowed) ...

3.9b ... and screw and nut at rear

3.9c Depress brake pedal while sprocket nut is slackened

3.12a Release metal battery holder (early models)...

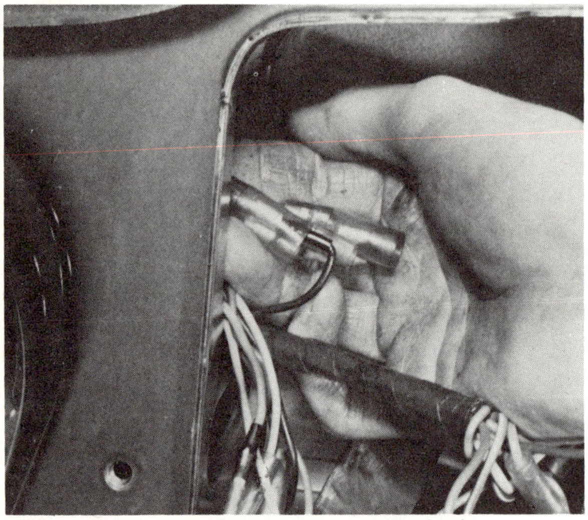

3.12b ... then trace and disconnect generator leads

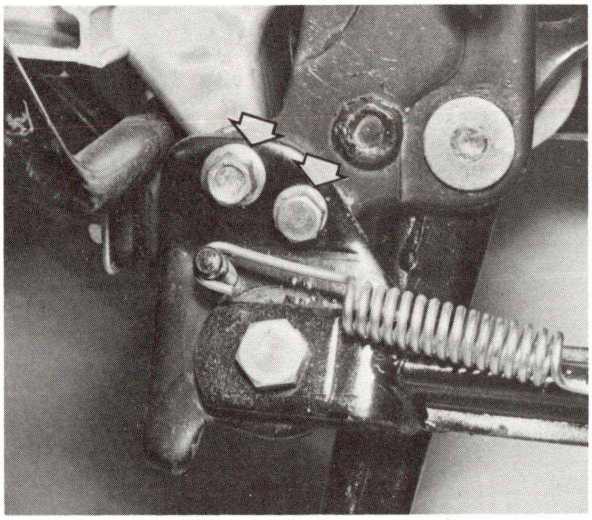

3.13 Footrest assembly is secured by these bolts

3.15 Front mounting bolt passes through air filter brackets

3.16 Lower unit slightly, and release neutral lead

4 Dismantling the engine, clutch and gearbox: general

Before commencing work on the engine unit, the external surfaces should be cleaned thoroughly. A motorcycle engine has very little protection from road grit and other foreign matter, which will find its way into the dismantled engine if this simple precaution is not observed. One of the proprietary cleaning compounds such as 'Gunk' can be used to good effect, particularly if the compound is allowed to work into the film of oil and grease before it is washed away. When washing down, make sure that water cannot enter the induction port or the electrical system, particularly if these parts have been exposed.

Never use undue force to remove any stubborn parts, unless mention is made of this requirement. There is invariably good reason why a part is difficult to remove, often because the dismantling operation has been tackled in the wrong sequence. Dismantling will be made easier if a simple engine stand is constructed that will correspond with the engine mounting points. This arrangement will permit the complete unit to be clamped rigidly to the workbench, leaving both hands free.

5 Preventing the engine from turning, both for dismantling and reassembly purposes

1 It is often necessary to stop the engine from rotating so that a component can be removed or tightened eg engine sprocket nut or clutch centre nut. One way of achieving this, which can be used during dismantling and reassembly, is by placing a round metal bar through the small end boss and resting this bar on two pieces of wood placed on top of the crankcase mouth. On no account must the metal bar be allowed to bear directly down onto the gasket face of the crankcase mouth, otherwise damage may occur causing a loss of primary compression.

2 Obviously, this method cannot be applied if the cylinder head and barrel are still in position, and it will therefore be necessary to adopt a different approach. One method is to select top gear and either apply the rear brake, or if the engine has been removed, wrap the rear chain around the gearbox sprocket so that it lodges against the engine casing.

3 A strap wrench or chain wrench may also prove invaluable, and these tools are available from most tool and accessory shops. It is fairly easy to improvise a chain wrench using an old drive chain and a suitable lever, as shown in photograph 5.3.

5.3 Chain wrench can be improvised as shown

6 Cylinder head: removal

1 The cylinder head is retained by four nuts, and may be removed easily, irrespective of whether or not the engine unit is installed in the frame. Slacken the four nuts diagonally and evenly to avoid any risk of distortion. Remove the nuts and spring washers, then lift the head away from the holding down studs. Check that the cylinder head gasket comes away with one or other of the mating surfaces, and that half of the gasket is not stuck to each. If the gasket is in good condition, it may be re-used.

7 Cylinder barrel: removal

1 The cylinder barrel may be drawn off the holding down studs, after the cylinder head has been removed. Some initial resistance may be encountered as the gasket compound used during manufacture tends to be tenacious. To help break the joint, the base of the barrel can be tapped using a block of hardwood and a hammer – never use a hammer directly, as the casting is easily fractured.

2 As the barrel is pulled up the holding down studs, stuff some rag into the crankcase mouth to prevent the ingress of debris. Support the connecting rod and piston as they emerge from the cylinder bore.

8 Piston and small end: removal

1 Remove both of the circlips retaining the gudgeon pin and discard them. They must never be reused.

2 Push out the gudgeon pin from the piston and release the piston from the connecting rod.

3 If the gudgeon pin is tight, warm the piston by wrapping it in a rag, soaked in boiling water and wrung out. Never drift the gudgeon pin out unless absolutely necessary, and only then if the piston is well supported, otherwise there is risk of bending the connecting rod.

4 Note that the piston crown is marked with an arrow and when reassembling, this arrow must point forwards, towards the exhaust port.

5 Push out the small end needle roller bearing.

6 Remove the cylinder barrel gasket and discard it.

9 Oil pump: removal

1 **Note:** It is not necessary to remove the oil pump for general dismantling purposes. See Section 11 for details about removing the right-hand inner cover with the oil pump undisturbed.

The oil pump is mounted behind the right-hand engine casing, to the rear of the carburettor. The pump can be removed with the engine installed in the frame, or after removal. In the case of the former it will be necessary to remove the outer cover and to release the pump control cable and feed pipe. Note that when the pipe is disconnected it will be necessary either to drain any oil into a clean container for re-use, or to plug the tank outlet. It will also be necessary to disconnect the oil feed pipe which runs from the front of the pump body to the crankcase, immediately below the carburettor stub. The pipe terminates in a banjo union which is secured by a single bolt.

2 The pump can be removed after the two securing screws have been removed. It may help to twist the pump body slightly as it is withdrawn, to aid disengagement of the driving worm gear. Note that it is essential that the oil pump is bled after installation, referring to Chapter 2, Section 13 for details.

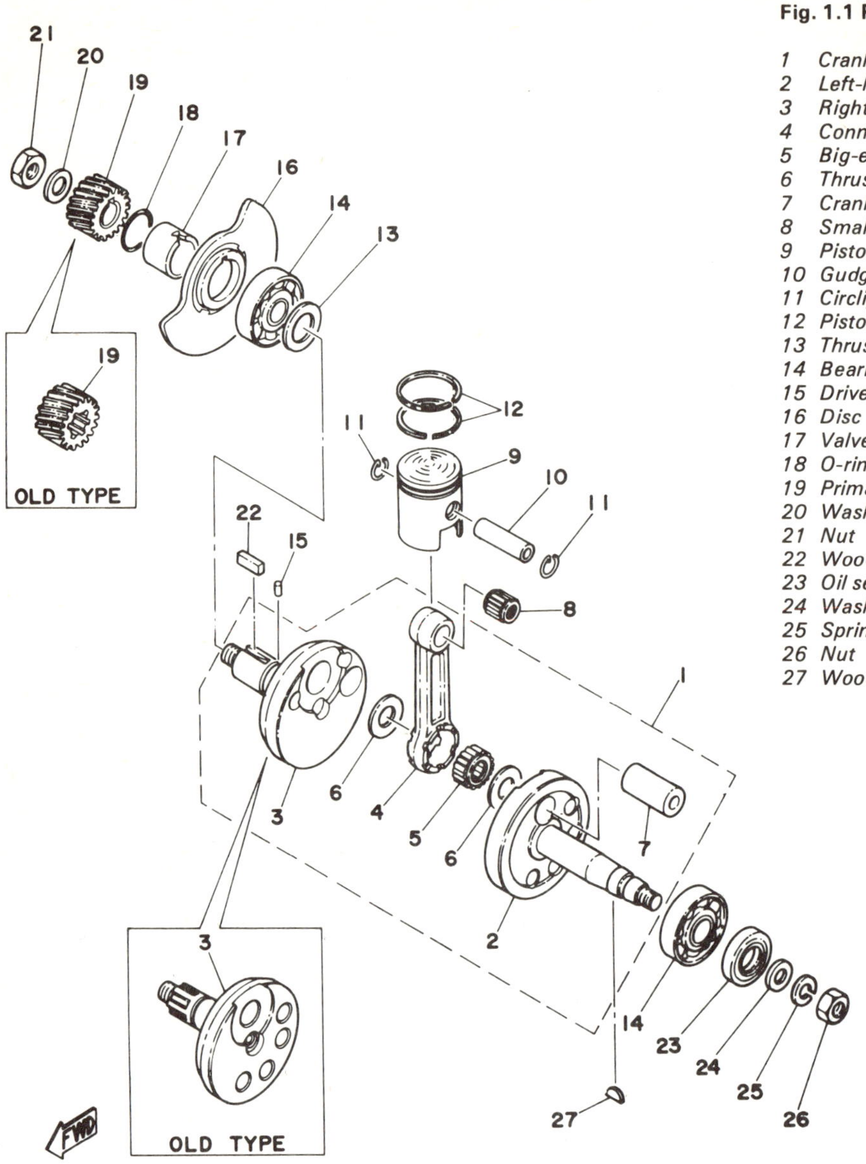

Fig. 1.1 Piston and crankshaft assembly

1 *Crankshaft assembly*
2 *Left-hand flywheel*
3 *Right-hand flywheel*
4 *Connecting rod*
5 *Big-end bearing*
6 *Thrust washer – 2 off*
7 *Crankpin*
8 *Small-end bearing*
9 *Piston*
10 *Gudgeon pin*
11 *Circlip – 2 off*
12 *Piston ring set*
13 *Thrust washer*
14 *Bearing – 2 off*
15 *Drive pin*
16 *Disc valve*
17 *Valve collar*
18 *O-ring*
19 *Primary drive pinion*
20 *Washer*
21 *Nut*
22 *Woodruff key*
23 *Oil seal*
24 *Washer*
25 *Spring washer*
26 *Nut*
27 *Woodruff key*

10 Gearbox sprocket: removal

1 The gearbox sprocket is retained to the splined mainshaft end by a large nut, the removal of which will necessitate locking the sprocket. With the engine in the frame, this is easily accomplished by applying the rear brake, thus immobilising the sprocket via the final drive chain. With the engine on the bench, a chain wrench, or the drive chain bunched against the casing, can be used, or top gear selected and a locking bar passed through the small end eye.

2 Having locked the sprocket in position, knock back the locking tab with a cold chisel, then unscrew the retaining nut. The sprocket and spacer may now be drawn off the mainshaft splines.

11 Right-hand inner cover: removal

1 The right-hand inner cover contains the kickstart mechanism, clutch assembly and disc valve assembly, and must be removed to gain access to these components. Before the cover is released, it is necessary to remove the carburettor. The oil pump and the clutch release lever will not impede removal, and can be left in position.

2 The cover is secured by the usual array of cross-head screws. These are usually very tight and may require the use of an impact driver to dislodge them. It should be noted that the normal impact driver bits may not be long enough to reach the more deeply recessed screws, and an extended bit should be used, where necessary.

3 With the casing screws removed, tap around the joint with a soft-faced mallet to break the seal, then draw the cover away. If the kickstart shaft oil seal is to be reused, take care not to damage it on the splines. It is a good idea to wrap some PVC tape around the splines to prevent seal damage, and this applies equally when the time comes for reassembly. Note that a small amount of residual oil may be released as the cover is removed, and some provision should be made to catch this.

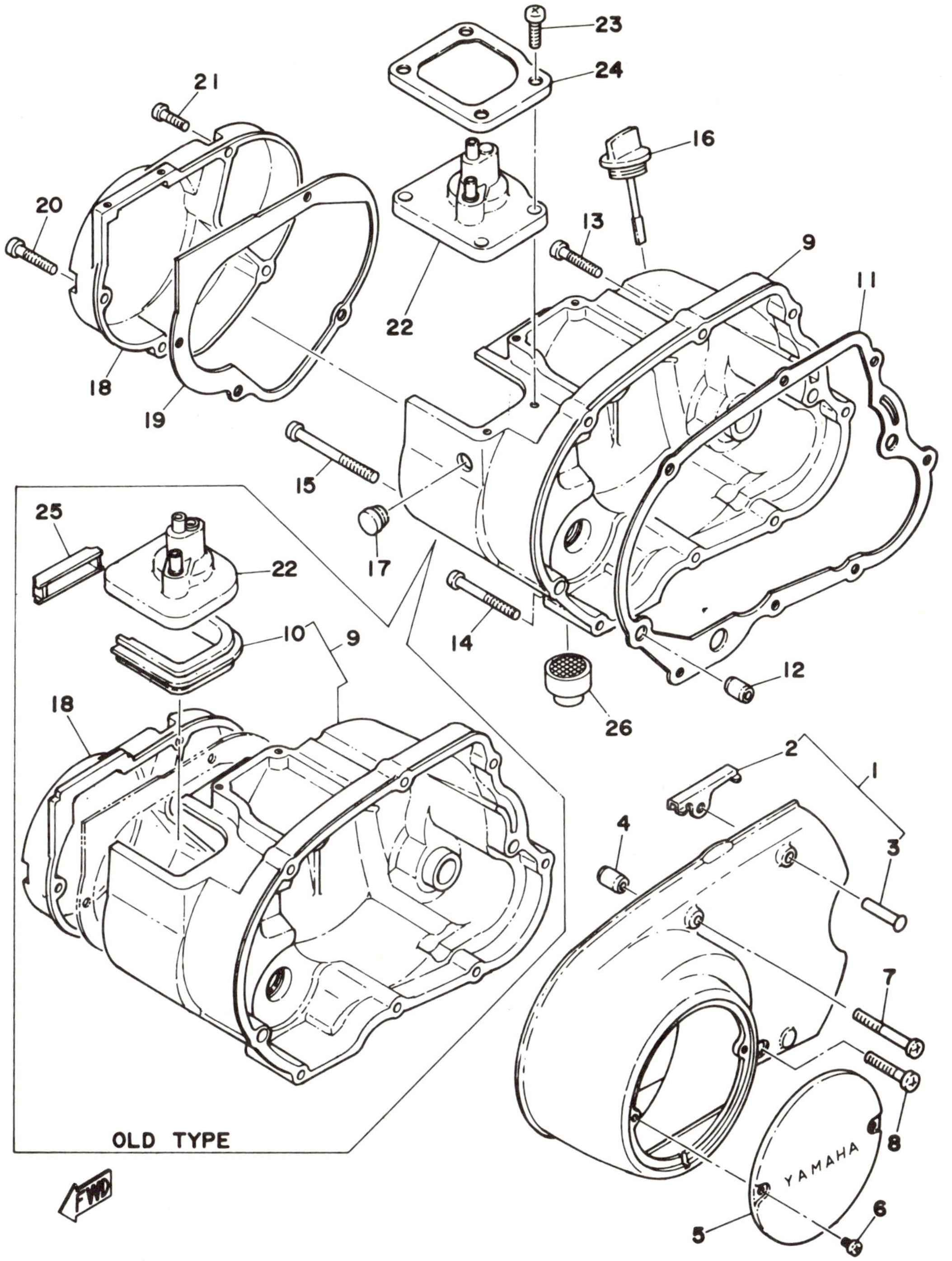

Fig. 1.2 Engine covers – early models

1 *Left-hand crankcase cover assembly*
2 *Guard plate*
3 *Rivet*
4 *Hollow dowel – 2 off*
5 *Generator cover*
6 *Screw – 2 off*
7 *Screw – 2 off*
8 *Screw*
9 *Right-hand crankcase cover assembly*
10 *Carburettor cover seal*
11 *Gasket for right-hand cover*
12 *Hollow dowel – 2 off*
13 *Screw – 6 off*
14 *Screw*
15 *Screw*
16 *Oil level plug*
17 *Grommet plug*
18 *Carburettor cover*
19 *Gasket for carburettor cover*
20 *Screw – 3 off*
21 *Screw*
22 *Carburettor cover*
23 *Screw – 4 off*
24 *Plate*
25 *Guide*
26 *Drain tube*

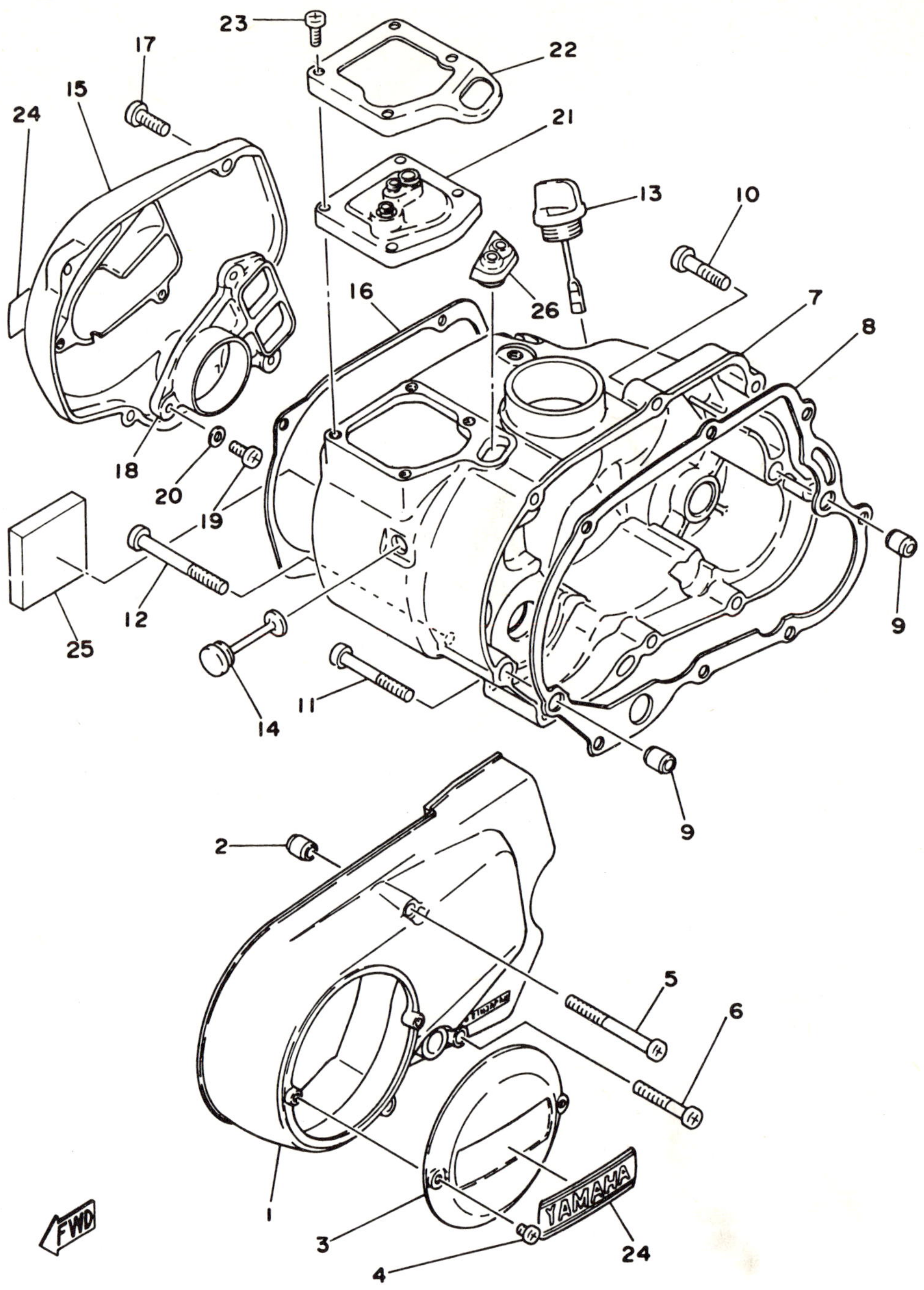

Fig. 1.3 Engine covers – later type

1 *Left-hand crankcase cover*
2 *Hollow dowel – 2 off*
3 *Generator cover*
4 *Screw – 2 off*
5 *Screw – 2 off*
6 *Screw*
7 *Right-hand crankcase cover*
8 *Gasket for right-hand cover*
9 *Hollow dowel – 2 off*
10 *Screw – 6 off*
11 *Screw*
12 *Screw*
13 *Oil level plug*
14 *Grommet plug*
15 *Carburettor cover*
16 *Gasket for carburettor cover*
17 *Screw – 4 off*
18 *Adaptor plate*
19 *Screw – 3 off*
20 *Washer – 3 off*
21 *Carburettor rubber cover*
22 *Carburettor cover retaining plate*
23 *Screw – 4 off*
24 *Decal*
25 *Drain*
26 *Oil pipe holder*

12 Clutch: removal

1 Before commencing any dismantling work, it should be noted that unless the cylinder head and barrel have been removed, the crankshaft pinion nut should be released before the clutch is disturbed, as removal at a later stage will prove to be difficult. To prevent crankshaft rotation, select top gear and apply the rear brake. Alternatively, if the unit is removed from the frame, a thick wad of rag can be wedged between the primary gear teeth.
2 Slacken in a diagonal sequence the six screws which retain the clutch pressure plate, then lift the pressure plate and screws away. If the unit is in position in the frame, select top gear and apply the rear brake so that the clutch centre can be held whilst the securing nut is slackened. If the engine unit is being dismantled on the bench, some method of restraining the clutch centre must be improvised. It was found that this could be accomplished by temporarily refitting two or three of the clutch springs, followed by some large washers and the clutch screws. This applies sufficient pressure on the clutch to enable it to hold whilst the centre nut is released. The accompanying photograph illustrates the method used.
3 Pass a stout bar through the connecting rod small end eye, arranging it to bear against wooden blocks at the crankcase mouth. The crankshaft, primary drive and clutch are now effectively locked and the securing nut may be released after knocking back the locking tab.
4 Remove the remaining screws and springs where appropriate, then lift out the clutch plain and friction plates as a group together with the clutch centre. It should be noted that two types of clutch have been used, these differing slightly in their construction. On the original type, the outer lip of the clutch centre is followed by a thick plain plate. Friction and thin plain plates are fitted next, followed by the inner pressure plate which can be identified by its six protruding bosses. Later clutches are similar in general design, but differ in that the thick outer plain plate is integral with the clutch centre.
5 With the clutch centre and plates removed, slide off the plain washer and thrust race, where fitted. Later machines make use of a plain thrust washer in place of this bearing. Lift away the clutch drum from the end of the gearbox mainshaft. On early machines, the kickstart driven pinion will remain on the shaft, and this should be removed together with its thrust washer. On later units, this pinion is integral with the clutch drum.

12.2 Lock clutch as shown, then release nut

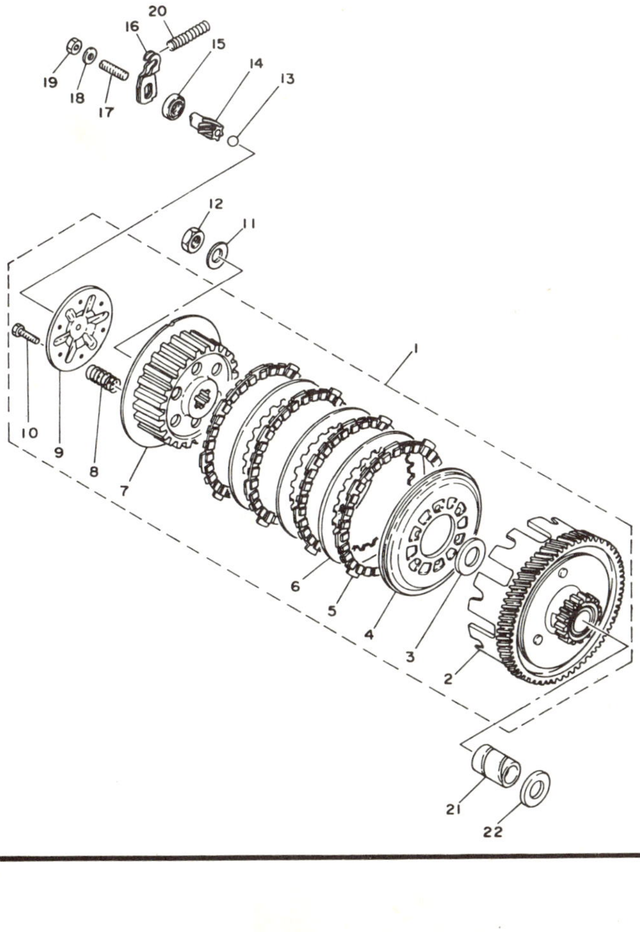

Fig. 1.4 Clutch – later models

1 Complete clutch assembly
2 Primary driven pinion and clutch outer drum (74 teeth)
3 Washer
4 Clutch pressure plate
5 Clutch friction plate – 4 off
6 Clutch plate – 3 off
7 Clutch centre
8 Spring – 6 off
9 Clutch pressure plate
10 Screw – 6 off
11 Washer
12 Nut
13 Ball bearing ($\frac{1}{4}$ inch)
14 Quick thread worm
15 Oil seal
16 Actuating lever
17 Adjusting screw
18 Washer
19 Nut
20 Spring
21 Spacer
22 Washer

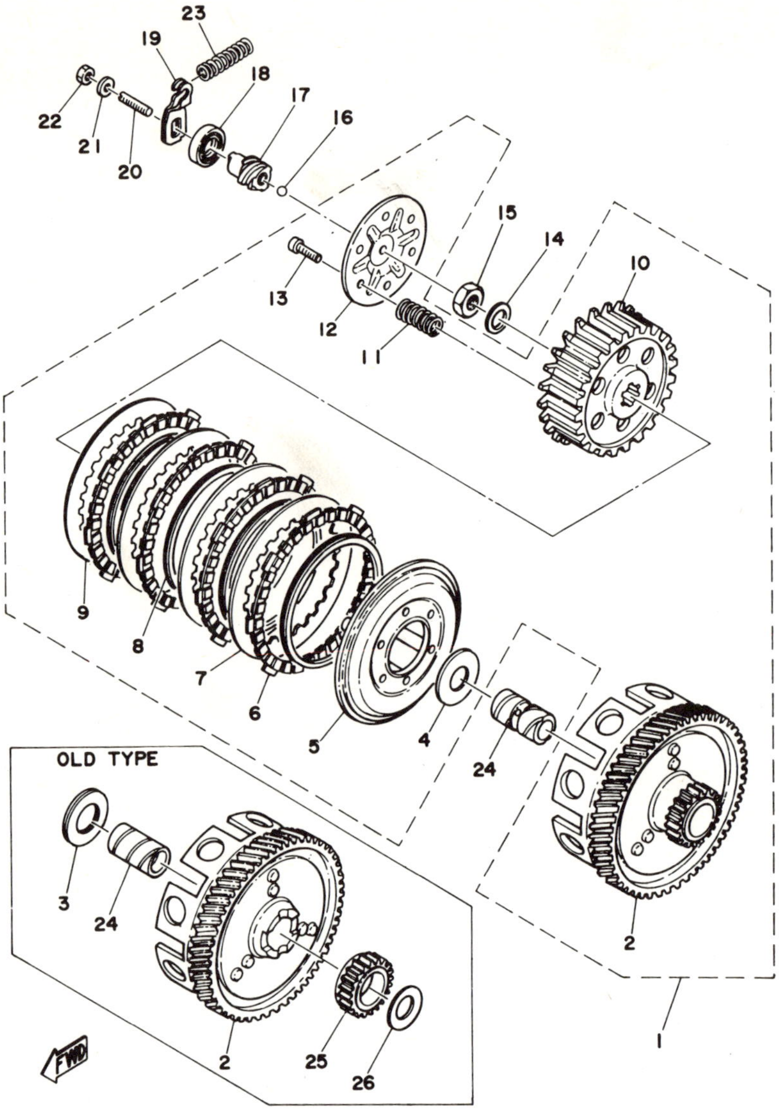

Fig. 1.5 Clutch assembly – early models

1 *Clutch assembly*
2 *Primary drive gear/clutch drum*
3 *Thrust bearing*
4 *Thrust washer*
5 *Clutch pressure plate*
6 *Friction plate – 4 off*
7 *Plain plate – 3 off*
8 *Cushion ring – 4 off*
9 *Outer plate*
10 *Clutch centre*
11 *Spring – 6 off*
12 *Pressure plate*
13 *Screw – 6 off*
14 *Washer*
15 *Lock nut*
16 *Ball bearing*
17 *Operating screw*
18 *Oil seal*
19 *Actuating lever*
20 *Adjusting screw*
21 *Washer*
22 *Nut*
23 *Return spring*
24 *Spacer*
25 *Kickstart pinion*
26 *Thrust plate*

13 Kickstart mechanism: removal

1 The kickstart shaft, spring and pinion form a self-contained unit, which can be lifted away without dismantling, after releasing the anchored end of the spring. The return spring is not under enormous pressure, and it will be found that the assembly can be twisted clockwise to unhook the spring end. Let the spring unwind slowly, then pull the assembly out of its recess in the casing.

2 With the assembly removed from the casing, it may be dismantled further if attention to this mechanism is required. On early models, release the circlip from the outer end of the kickstart shaft, then slide off the outer spring seat. The return spring may now be removed, as can the inner spring seat. The kickstart pinion runs on a quick thread machined into the shaft end, and is retained by a circlip and a thrust washer.

3 The kickstart mechanism on late models is arranged slightly differently (see Fig. 1.7). Slide the spacer off the kickstart shaft, followed by the spring itself. Remove the large plain washer. The pinion is retained by a split collet arrangement, the collet halves being held in plate by a large circlip which fits around a groove on their outer edge. Remove the circlip and displace the collet halves. The pinion can then be removed from the shaft.

4 The kickstart idler pinion is free-running on the protruding end of the gearbox layshaft. It can be removed after releasing the circlip which retains it. Note that a wave washer is fitted beneath the pinion, and this should be removed and placed with the pinion to preclude its loss.

14 Gear selector mechanism: removal

1 Movement from the gear change pedal is conveyed across the engine unit by way of a shaft. At the right-hand end, the shaft terminates in a pressed-steel selector mechanism. This arrangement takes the form of a claw which operates on the end of the selector drum by pushing or pulling against pins, thus turning the selector drum. A stopper arm and roller act as a detent mechanism which limits the amount of selector drum movement at each change.

2 Prise off the circlip which retains the cross-over shaft on the left-hand side of the unit. The selector claw should be lifted clear of the casing. Release the spring from the stopper arm, then slacken the shouldered bolt upon which it pivots. The stopper arm can now be lifted away, together with the spring.

13.1a Disengage kickstart return spring ...

13.1b ... then lift assembly out of casing

13.2a Remove upper spring seat

13.2b Disengage and remove return spring ...

13.2c ... followed by lower spring seat

13.2d Pinion is retained by a circlip

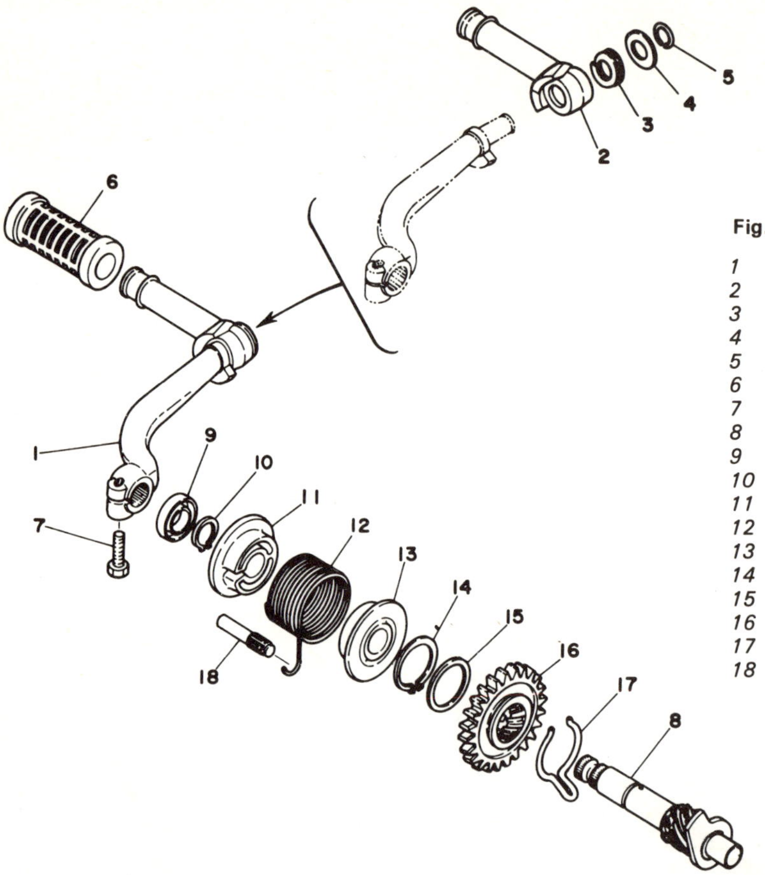

Fig. 1.6 Kickstart mechanism – early type

1 *Kickstart crank assembly*
2 *Kickstart lever*
3 *Kickstart crank spring*
4 *Kickstart lever washer*
5 *Kickstart lever clip*
6 *Kickstart lever rubber*
7 *Bolt for kickstart crank*
8 *Kickstart spindle assembly*
9 *Oil seal*
10 *Circlip*
11 *Kickstart return spring collar*
12 *Kickstart return spring*
13 *Kickstart return spring guide*
14 *Circlip*
15 *Shim*
16 *Kickstart drive pinion (26T)*
17 *Kickstart assembly clip*
18 *Kickstart spring anchor*

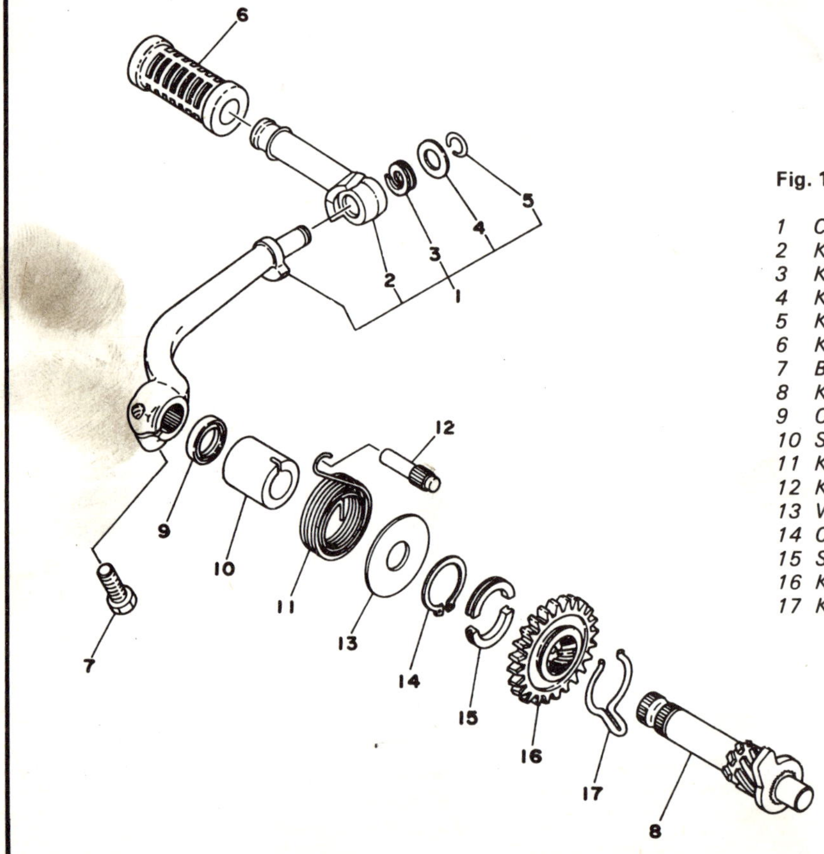

Fig. 1.7 Kickstart assembly – later type

1 *Complete kickstart crank assembly*
2 *Kickstart lever*
3 *Kickstart crank spring*
4 *Kickstart lever washer*
5 *Kickstart lever circlip*
6 *Kickstart lever rubber*
7 *Bolt*
8 *Kickstart spindle assembly*
9 *Oil seal*
10 *Spacer*
11 *Kickstart return spring*
12 *Kickstart spring anchor*
13 *Washer*
14 *Circlip*
15 *Shim*
16 *Kickstart drive pinion (26 teeth)*
17 *Kickstart assembly clip*

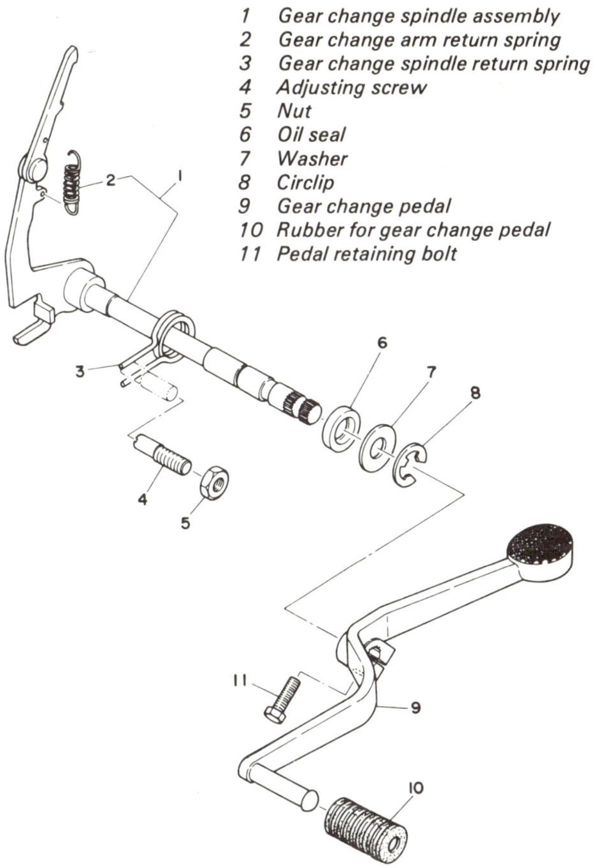

Fig. 1.8 Gear change mechanism

15 Disc valve assembly: removal

1 **Important note**: Before commencing removal, note that the valve disc must be refitted in its original position upon reassembly, or the engine will not run properly. Pay particular attention to the remarks on alignment given in this Section.
2 If the crankshaft pinion has not been removed at this stage, slacken the retaining nut and remove the washer and pinion. Crankshaft rotation may be prevented by passing a stout bar through the connecting rod small end eye, and allowing the ends to rest on wooden blocks arranged at each side of the crankcase mouth. Note that on early models, the crankshaft end and internal bore of the pinion are splined, whilst on later models a rectangular key is used to locate the pinion.
3 Slacken and remove the six screws which secure the valve cover. It may prove helpful to tap around the jointing face of the cover to break the seal. Lift the cover away, taking care not to disturb the components beneath.
4 With the cover removed, it will be possible to take note of the disposition of the various components. Start by setting the crankshaft at top dead centre (TDC). On early models, a steel sleeve fits over the protruding mainshaft. It is located at its inner end by a small pin which passes through a slot in the sleeve, and into the boss of the valve disc. It will be noted that, at TDC, the pin and slots are in line with the connecting rod, and that the valve disc cutaway is slightly offset in relation to the inlet port. It is recommended that a simple sketch is made of this arrangement as an aid to correct reassembly. Mark lightly the valve disc boss immediately adjacent to the pin to ensure that it is not inverted when reassembled.
5 On later machines, a splined insert is located on the crankshaft by the pin, the valve disc having corresponding splines. Note that index marks are provided on the insert and valve disc boss, and these should align with the driving pin when the assembly is installed correctly.
6 Having made note of the above, the valve disc and collar may be removed and placed with the cover to await further examination. The small driving pin can be removed by grasping it with a pair of pointed-nosed pliers and pulling it out of the crankshaft end. If the pin proves to be a tight fit, it can be driven out, with the aid of a small punch, from the opposite side. This can be improvised by using a masonary nail or a large pop rivet pin, these being suitably hardened.

16 Flywheel generator assembly: removal

1 The flywheel rotor is keyed to the left-hand end of the crankshaft, and is secured by a nut and spring washer. It will be necessary to prevent the crankshaft from turning whilst the nut is slackened, using one of the methods described in Section 5 of this Chapter.
2 It will be necessary to draw the rotor off the mainshaft end, using an extractor which screws into the thread in the outer face of the rotor. There is no safe alternative to this method, because there is insufficient clearance between the rotor edge and stator to employ a universal legged puller. It may be possible to borrow or hire this tool (Yamaha part number 90890–01189) from a local Yamaha Service Agent. In view of its necessity, when renewing the contact breaker assembly for instance, the extractor should be regarded as essential.
3 Screw the extractor into the rotor boss, then gradually tighten the centre bolt whilst the body of the extractor is held by means of an open-ended spanner on its flats. If the rotor does not draw off fairly easily, do not resort to applying excessive pressure to the centre bolt. Instead, use a hammer and punch to tap gently around the rotor boss. This will often succeed in jarring the rotor free of its taper joint.
4 Lift the rotor away, then displace the Woodruff key from its keyway in the crankshaft end. Slacken the two large countersunk screws which secure the generator stator assembly to the crankcase. Lift the stator away, disengaging the output leads from the casing recesses.

17 Separating the crankcase halves

1 Arrange the bare crankcase assembly on a workbench, using wooden blocks to support it, with the right-hand casing uppermost. The crankcase halves are retained by a total of twelve securing screws. These often prove very difficult to remove by normal means, as the heads are very easily damaged if sufficient pressure is applied by hand to loosen them.
2 The most practicable method is to use an impact driver, with an extended bit in some instances, which imparts a percussive twisting action. This will invariably jar the screw free without damaging the screw heads. These tools are not expensive, and can be considered almost essential for owners of Japanese machines.
3 Having removed the securing screws, the right-hand casing half can be lifted away, leaving the crankshaft and gearbox components in the left-hand casing half. It is likely that a certain amount of gentle persuasion will be necessary to effect separation, namely by tapping around the joint with a soft faced mallet, or a hammer and a block of hardwood.
4 Resist the temptation to lever the casing halves apart with a screwdriver, as this almost invariably leads to damaged sealing faces. It is particularly important that the crankcase joint remains absolutely airtight on a two-stroke engine, as an air leak can cause loss of primary compression, and can upset the fuel-air mixture.

5 If crankcase separation proves particularly difficult, it may prove expedient to take the bare crankcase assembly to a Yamaha Service Agent who will have the necessary service tool (number 90890-01135) with which to effect separation. It is normally possible to separate the crankcase halves without resorting to this method.

17.1 Crankcase is secured by twelve screws (arrowed)

18 Gearbox components and crankshaft: removal

1 Release the screws which secure the white plastic housing at the end of the gearbox selector drum. The housing can be lifted away, and the C-shaped retainer displaced to free the selector drum end. Unscrew the neutral switch assembly from the casing. The gearbox components can be removed as an assembly together with the selector drum and forks, and need not be disturbed unless specific attention is required. Place the assembly to one side.
2 The crankshaft assembly can be lifted out of the casing, and does not normally require much effort to free it from its bearing. If necessary, the protruding end of the crankshaft can be tapped with a soft-faced mallet to aid removal, but on no account should heavy blows be applied here, as it is possible to distort the crankshaft if excessive force is used.

19 Examination and renovation: general

1 Before examining the component parts of the dismantled engine/gear unit for wear, it is essential that they should be cleaned thoroughly. Use a paraffin/petrol mix to remove all traces of oil and sludge which may have accumulated within the engine.
2 Examine the crankcase castings for cracks or other signs of damage. If a crack is discovered, it will require professional attention, or in an extreme case, renewal of the casting.
3 Examine carefully each part to determine the extent of wear. If in doubt, check with the tolerance figures whenever they are quoted in the text. The following sections will indicate what type of wear can be expected and, in many cases, the acceptable limits.
4 Use clean, lint-free rags for cleaning and drying the various components, otherwise there is a risk of small particles obstructing the internal oilways.

20 Engine casings, bearings and oil seals: examination and renovation

1 The aluminium alloy casings and covers are unlikely to suffer damage through ordinary use. Damage can occur however if the machine is dropped, or if sudden mechanical breakages occur, such as the rear chain breaking.
2 Small cracks or holes may be repaired with an epoxy resin adhesive, such as Araldite, as a temporary expedient. Permanent repairs can only be effected by argon-arc welding, and a specialist in this process is in a position to advise on the viability of proposed repair. Often it may be cheaper to buy a new replacement.
3 Damaged threads can be economically reclaimed by using a diamond section wire insert, of the Helicoil type, which is easily fitted after drilling and re-tapping the affected thread. The process is quick and inexpensive, and does not require as much preparation and work as the older method of fitting brass, or similar inserts. Most motorcycle dealers and small engineering firms offer a service of this kind.
4 Sheared studs or screws can usually be removed with screw extractors, which consist of tapered, left-hand thread screws, of very hard steel. These are inserted by screwing anticlockwise into a pre-drilled hole in the stud, and usually succeed in dislodging the most stubborn stud or screw. The only alternative to this is spark erosion, but as this is a very limited, specialised facility, it will probably be unavailable to most owners. It is wise, however, to consult a professional engineering firm before condemning an otherwise sound casing. Many of these firms advertise regularly in the motorcycle papers.
5 The crankshaft main bearings and gearbox bearings should be examined for wear and roughness when turned, and if suspect, should be renewed. Remove the bearing oil seal retainer and prise out the old seal, where appropriate. The bearings can easily be removed by applying heat to the casing, causing the aluminium alloy to expand at a faster rate than that of the steel bearing, allowing the bearing to become oose. The safest way of doing this is to place the casing in an oven, heating it to about 80° – 100°C. The casing can then be banged on a wooden bench or board, face down, to jar the bearing free. The new bearings can be tapped into position using a large diameter socket as a drift. Care should always be exercised when heating alloy casings as excessive or localised heat can easily cause warpage. Seek specialist advice if you are not familiar with this task.
6 Main bearing failure will immediately be obvious when the bearings are inspected, after the old oil has been washed out. If any play is evident or if the bearings do not run freely, renewal is essential. Warning of main bearing failure is usually given by a characteristic rumble that can readily be heard when the engine is running. Some vibration will also be felt, which is transmitted via the footrests.
7 Oil seal failure is a common occurrence in two-stroke engines that have seen a reasonable amount of service. When the oil seals begin to wear, air is admitted to the crankcase which will dilute the incoming mixture. This in turn causes uneven running and difficulty in starting.
8 Examine the seals carefully, paying particular attention to the thin lip of each seal. This area performs the sealing function, and the seal should be renewed without question if it is scored or marked in any way. In view of the important part these seals play, it is considered good practice to renew them as a matter of course whilst the engine is stripped for rebuilding. A worn seal can be prised out of position without having to remove the relevant bearing. When fitting a new seal, take great care not to damage or distort it. Tap the seal gently into place, using a large socket or similar to ensure that it is fitted squarely.
9 Bushes may be dealt with in a similar manner to that described for ball bearings. Check the fit of the relevant shaft end in the bush. It should be a light fit, without any discernible free play.
10 It will be noted that the left-hand end of the gearbox

mainshaft is supported by a needle roller bearing, and this is fitted in a blind bore in the casing. If renewal is required, it will be necessary to use a bearing extractor or a slide hammer as shown in photograph 20.10.

21 Crankshaft assembly: examination and renovation

1 The crankshaft assembly comprises two full flywheels, two mainshafts, a crankpin and big end bearing, a connecting rod and a caged needle roller small end bearing. The general condition of the big end bearing may be established with the assembly removed from the engine, or with just the cylinder head and barrel removed, as would be the case during a normal decoke. In this way it is possible to decide whether big end renewal is necessary, without a great deal of exploratory dismantling.

2 Big end failure is characterised by a pronounced knock which will be most noticeable when the engine is worked hard. The usual causes of failure are normal wear, or a failure of the lubrication supply. In the case of the latter, big end wear will become apparent very suddenly, and will rapidly worsen. Check for wear with the crankshaft set in the TDC (top dead centre) position, by pushing and pulling the connecting rod,. No discernible movement will be evident in an unworn bearing, but care must be taken not to confuse end float, which is normal, and bearing wear.

3 If play is found or suspected, it is recommended that the complete crankshaft assembly is taken to a Yamaha Service Agent, who will be able to confirm the worst, and supply a new or service-exchange assembly. The task of dismantling and reconditioning the big-end assembly is a specialist task, and is considered to be beyond the scope and facilities of the average owner.

4 The small end bearing is of the caged needle roller type, and will seldom give trouble unless a lubrication failure has occurred. The gudgeon pin should be a good sliding fit in the bearing without any play. The bearing must be tested whilst it is in place in the small end eye. If play develops, a noticeable rattle will be heard when the engine is running, indicative of the need for bearing renewal.

5 No problem is encountered when replacing the caged needle roller bearing as it is a light push fit in the eye of the connecting rod. New small end bearings are normally supplied whenever the crankshaft assembly is renewed or service-exchanged.

Fig. 1.9 Cylinder head and crankcase

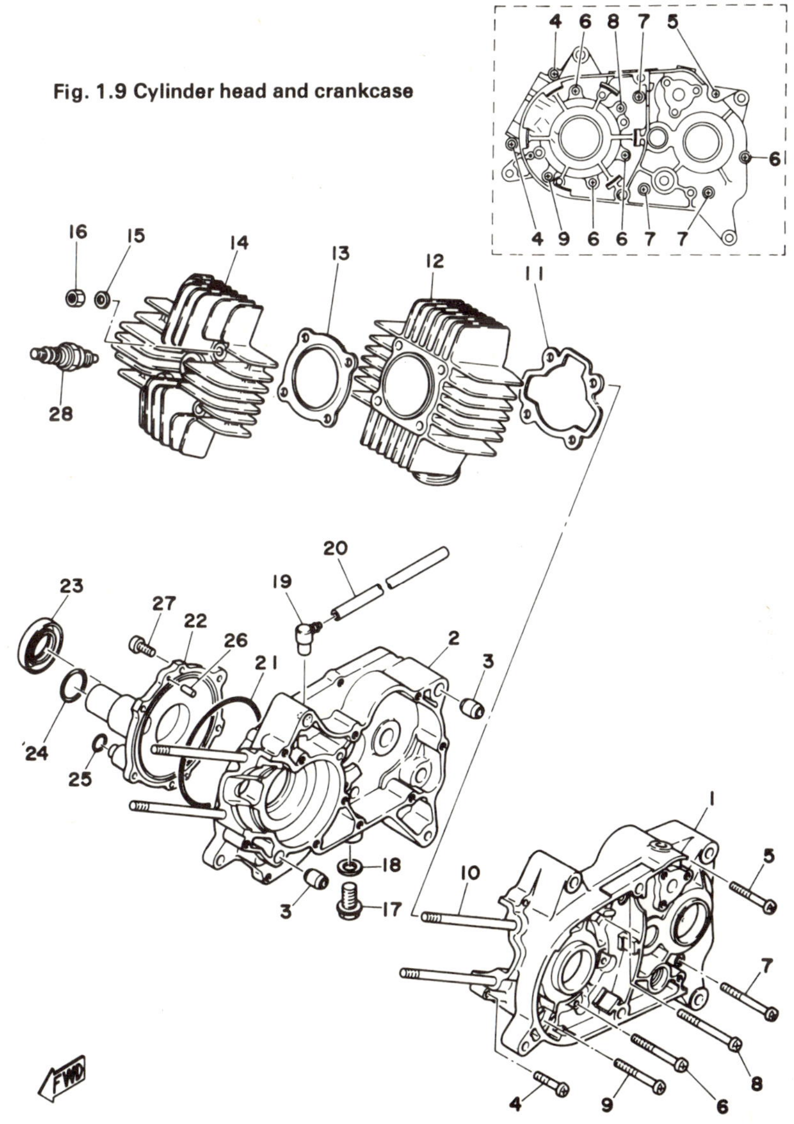

1 Left-hand crankcase half
2 Right-hand crankcase half
3 Hollow dowel – 2 off
4 Screw – 2 off
5 Screw
6 Screw – 4 off
7 Screw – 3 off
8 Screw
9 Screw
10 Stud – 4 off
11 Cylinder base gasket
12 Cylinder barrel
13 Cylinder head gasket
14 Cylinder head
15 Plain washer – 4 off
16 Nut – 4 off
17 Drain plug
18 Drain plug washer
19 Breather plug
20 Breather pipe
21 O-ring
22 Disc valve cover
23 Oil seal
24 O-ring
25 O-ring
26 Dowel pin – 2 off
27 Screw – 6 off
28 Sparking plug

20.5a Clean all bearings and feel for wear

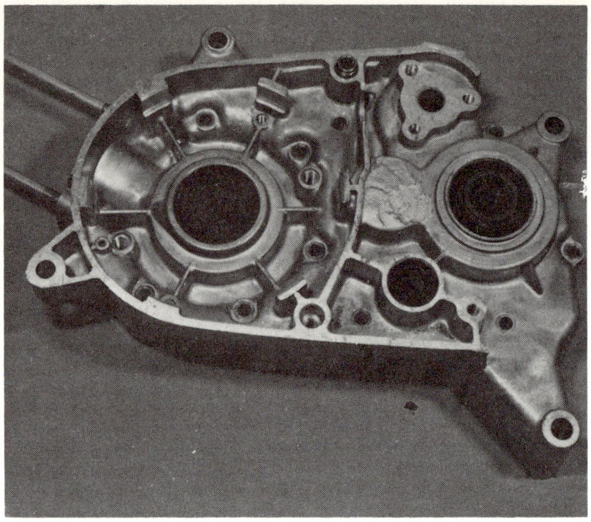

20.5b Oil seals may be removed with bearings in place

20.5c Note locating lips for some bearings

20.5d Check castings carefully for cracks or damage

20.10 Slide hammer or similar was needed here

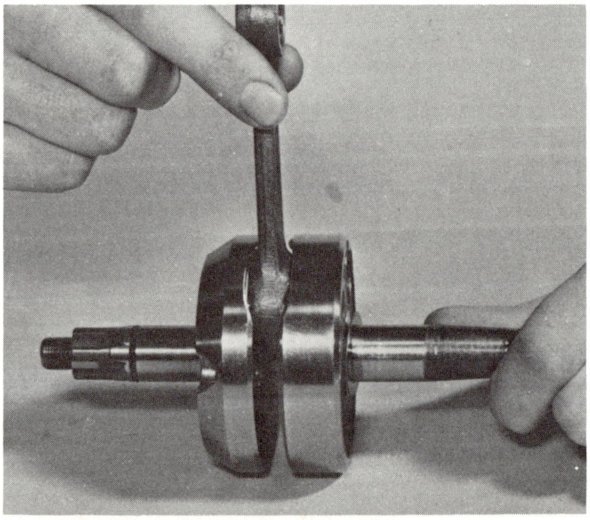

21.2 Check big end bearing for play at TDC

22 Decarbonising

1 Decarbonising must take place as part of any major overhaul, in addition to being a normal routine maintenance function. In the case of the latter, the operation can be undertaken with minimal dismantling, namely removal of the cylinder head. Carbon build up in a two-stroke engine is more rapid than that of its four-stroke counterpart, due to the oily nature of the combustion mixture. It is however, rather softer and is therefore more easily removed.
2 The object of the exercise is to remove all traces of carbon whilst avoiding the removal of the metal surface on which it is deposited. It follows that care must be taken when dealing with the relatively soft alloy cylinder head and piston. Never use a steel scraper or screwdriver for carbon removal. A hardwood, brass or aluminium scraper is the ideal tool as these are harder than the carbon, but no harder than the underlying metal. Once the bulk of the carbon has been removed, a brass wire brush of the type used to clean suede shoes can be used to good effect.
3 The whole of the combustion chamber should be cleaned, as should the piston crown. It is recommended that as smooth a finish as possible is obtained, as this will slow the subsequent build up of carbon. If desired metal polish can be used to obtain a smooth surface. The exhaust port must also be cleaned out, as a build up of carbon in this area will restrict the flow of exhaust gases from the cylinder. Take care to remove all traces of debris from the cylinder and ports, prior to assembly.

23 Cylinder head: examination and renovation

1 Using a wire brush, clean out any road dirt or other debris from the cylinder head fins, to prevent any possibility of overheating.
2 Check the condition of the thread in the sparking plug hole. If it is damaged an effective repair can be made using a Helicoil thread insert. This service is available from most Yamaha Service Agents. The cause of a damaged thread can usually be traced to overtightening of the plug or using a plug of too long a reach. Always use the correct plug and do not overtighten.
3 Check the cylinder head for warpage (usually caused by uneven tightening and/or overtightening), with a straight edge across several places on the gasket face; or preferably, with engineers' blue on a surface plate (a sheet of plate glass can be used as a substitute for a surface plate). If the cylinder head is warped, grind it down on a surface plate with emery paper. Start with 200 grade paper and finish with 400 grade and oil.
4 If it is necessary to remove a substantial amount of metal before the cylinder head will seat correctly, a new cylinder head should be obtained.

24 Cylinder barrel and bore: examination and renovation

1 Clean the outside of the cylinder barrel, taking care to remove any accumulation of dirt from between the cooling fins. Carefully remove the ring of carbon from the mouth of the bore, so that an accurate assessment of bore wear can be made.
2 A close visual examination of the bore surface must be made, to check for scoring or any other damage, particularly if broken piston rings were encountered during the stripdown. Any damage of this nature will necessitate reboring and a new piston, as it is impossible to obtain a satisfactory seal if the bore is not perfectly finished.
3 There will probably be a lip at the uppermost end of the cylinder bore which marks the limit of travel of the top of the piston ring. The depth of the lip will give some indication of the amount of bore wear that has taken place even though the amount of wear is not evenly distributed.
4 The best way of measuring bore wear is by the use of a cylinder bore DTI (Dial Test Indicator) or a bore micrometer. However, it is most unlikely that the average owner will have this type of equipment at his disposal. A slightly less accurate but more practical method is to insert the piston into the cylinder bore, and measure the gap between the piston skirt and bore using feeler gauges. The measurement should be taken at various positions and the average clearance assessed. It should be noted that the curvature of the gap precludes accurate measurement by this method, but it should be possible to gain a good indication of whether a rebore is necessary by comparing the measurements taken at an unworn and worn areas of the bore surface.
5 If the amount of wear exceeds 0.05 mm (0.002 in) it will be necessary to have the cylinder barrel rebored to the next oversize, and the ariate oversize piston fitted. Care must be exercised if the bore proves to be part-worn (ie 0.03 mm) and new rings are required, as a ridge will be present at the top of the bore, and the new top ring may strike the ridge, causing it to fracture. It is suggested that the advice of a Yamaha Service Agent be sought if this likelihood arises. Note that bore wear must always be assessed in conjunction with piston and piston ring conditions as described in the following Section.

25 Piston and piston rings: examination and renovation

1 Remove the piston rings from the piston by placing a thumb at each end, and gently expanding the ring until it can be slid off the piston. Great care must be taken, as the rings are very brittle and will shatter if stretched too far. If inexperienced in piston ring removal, it is as well to use feeler gauges as shown in the accompanying diagram. Take care that the rings are kept separate and the right way up so that they may be replaced in the correct position.
2 Check the piston rings by placing each ring in the bottom of the bore (this is the least worn part of the bore). Press it down a little way with the piston to make sure that it is square in the bore. Measure the end gap and compare with the wear limit. Renew if necessary.
3 When refitting new rings always check the end gap and enlarge it, if necessary, by filing with a needle file.
4 Check that the piston and bore are not scored, particularly if the engine has tightened up or seized. If the bore is badly scored, it will require a rebore and oversize piston. If the scoring is not too severe or the piston has just picked up, it is possible to remove the high spots by careful use of a needle file. Do not try to remove the file marks completely, since they will act as oil pockets and assist during the initial bedding in.
5 Before replacing the rings on the piston, make sure that the ring lands are clear of carbon. Be very careful not to damage the lands when cleaning. Also check that the ring locating pegs are not worn. (If they are, a new piston will have to be obtained).
6 The rings are identical when new, but after use will have worn and thus must be replaced in the same piston groove. Ensure that the ring is the correct way up ie; stamped mark facing upwards and that the gaps are positioned over the locating pins.
7 Examine the gudgeon pin for scores or stepped wear, and renew if necessary. Check the gudgeon pin to piston fit, renew if necessary.
8 Note that when new rings are fitted to a used bore, the surface glaze of the bore must be removed to enable the rings to bed in to the bore surface. This is best done with a glaze breaking tool, but may be done by hand, using medium grade emery cloth. No significant amount of metal should be removed; it is necessary only to lightly score the surface to remove the polished finish.

26 Piston/bore clearance check

1 Measure the piston diameter 10 mm (0.394 in) above the piston skirt, perpendicular to the gudgeon pin hole.
2 Measure the cylinder bore using an inside micrometer and substract from this figure the piston diameter obtained in

paragraph 1. Compare the result with the Specifications and renew if necessary.
3 Alternatively, if the appropriate measuring equipment is not available, an approximate check of the piston and its bore clearance can be made, using a feeler gauge.

27 Gearbox components: examination and renovation

1 Examine each of the gear pinions to ensure that there are no chipped or broken teeth and that the dogs on the end of the pinions are not rounded. Gear pinions with these defects must be renewed, there is no satisfactory method of reclaiming them.
2 Examine the selector forks carefully, ensuring that there is no scoring or wear where they engage in the gears, and that they are not bent. Damage and wear rarely occur in a gearbox which has been properly used and correctly lubricated, unless very high mileages have been covered.
3 The tracks in the selector drum, which co-ordinate the movement of the selector forks, should not show signs of undue wear.
4 Check the gear selector forks/gear pinion groove clearance, renew as necessary.
5 Check that the gear selector forks are not bent or cracked (particularly near the webbing).
6 Check the condition of the kickstart ratchet teeth. If they are worn and rounded, the kickstart will slip under load.

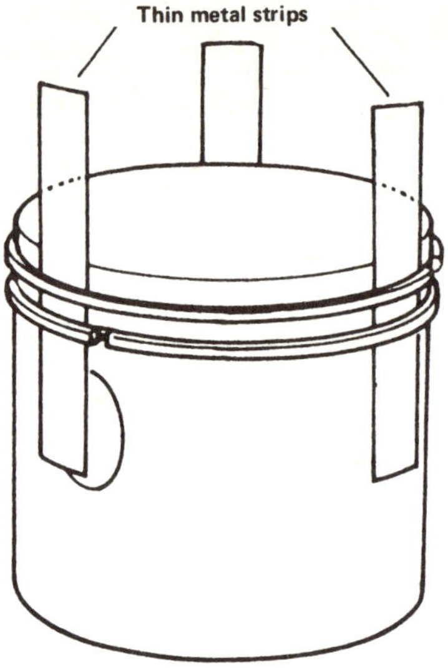

Fig. 1.10 Freeing gummed rings

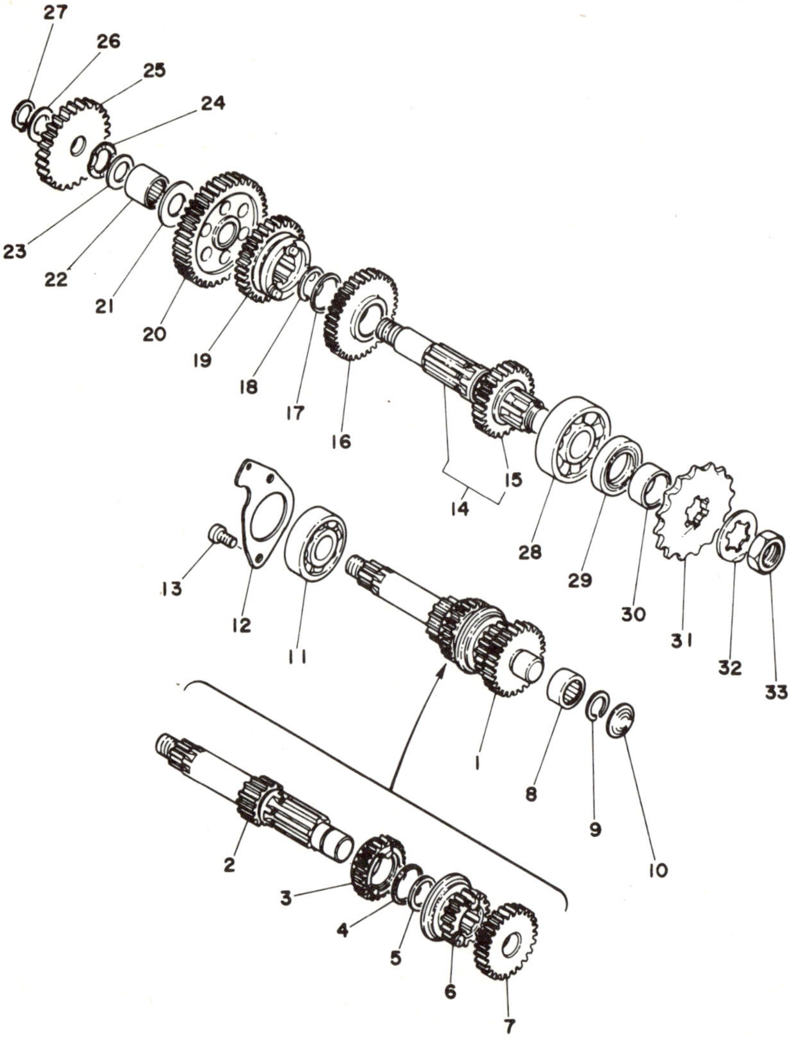

Fig. 1.11 Gearbox components

1 Complete mainshaft
2 Mainshaft
3 Mainshaft 3rd gear pinion (23T)
4 Thrust washer
5 Circlip
6 Mainshaft 2nd gear pinion (18T)
7 Mainshaft 4th gear pinion (27T)
8 Bearing
9 Circlip
10 Cap
11 Bearing
12 Bearing cover plate
13 Screw – 2 off
14 Complete layshaft
15 Layshaft 4th gear pinion (26T)
16 Layshaft 2nd gear pinion
17 Thrust washer
18 Circlip
19 Layshaft 3rd gear pinion (30T)
20 Layshaft 1st gear pinion (40T)
21 Shim washer
22 Needle roller bearing
23 Thrust washer
24 Wave washer
25 Kickstarter idler pinion
26 Thrust washer
27 Circlip
28 Bearing
29 Oil seal
30 Distance collar
31 Final drive sprocket
32 Lock washer
33 Lock nut

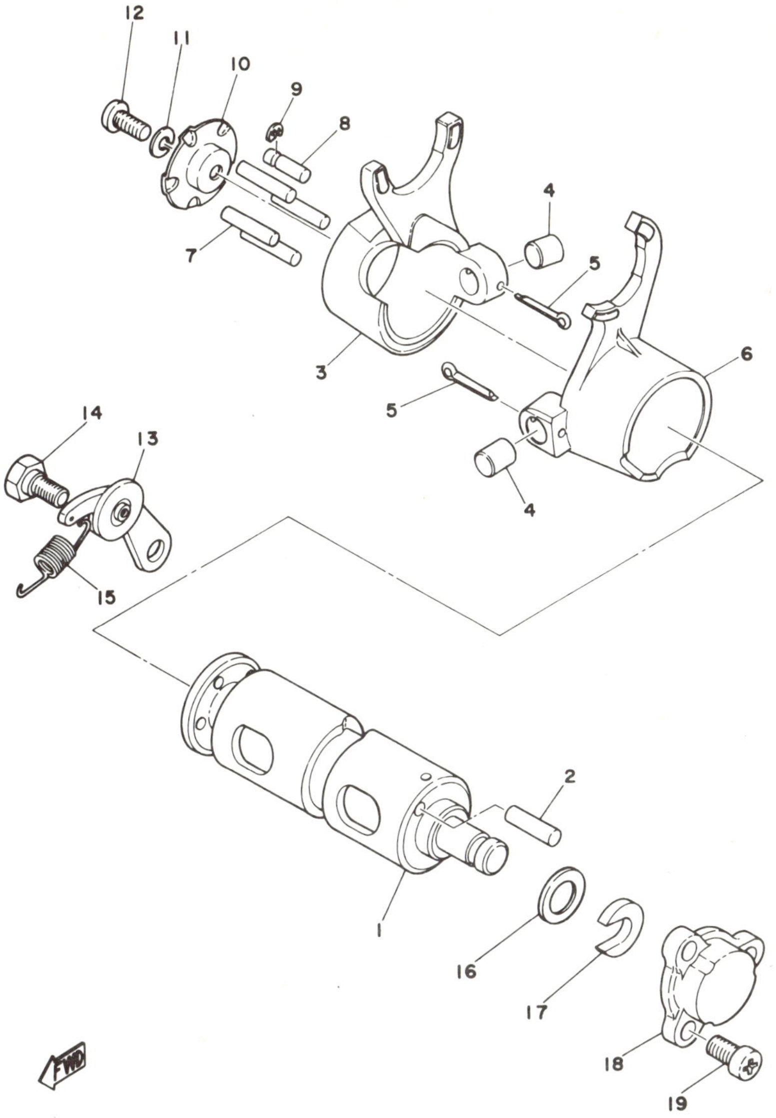

Fig. 1.12 Gear selector drum

1 *Cam*
2 *Dowel pin*
3 *Gear selector fork (shift 2)*
4 *Cam follower pin – 2 off*
5 *Split pin – 2 off*
6 *Gear selector fork (shift 1)*
7 *Dowel pin – 4 off*
8 *Locating pin*
9 *Circlip*
10 *Side plate*
11 *Spring washer*
12 *Screw*
13 *Stop lever assembly*
14 *Screw*
15 *Stop spring*
16 *Plain washer*
17 *Retainer*
18 *Blanking off plug*
19 *Screw – 3 off*

27.2a Selector pins are unlikely to show any wear

27.2b Raised pip operates neutral light switch

27.2c Release split pin and ...

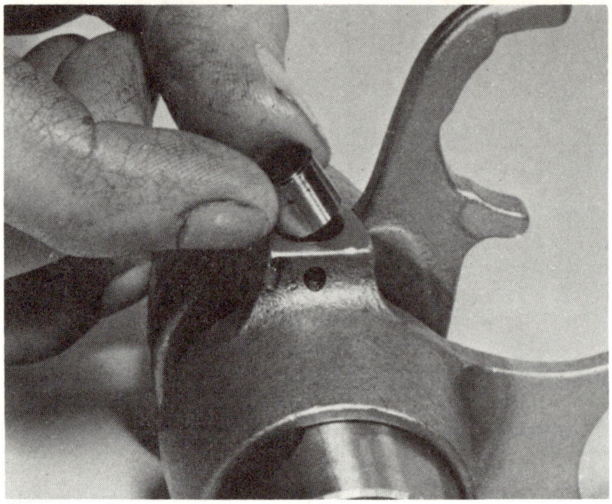
27.2d ... displace selector cam pin ...

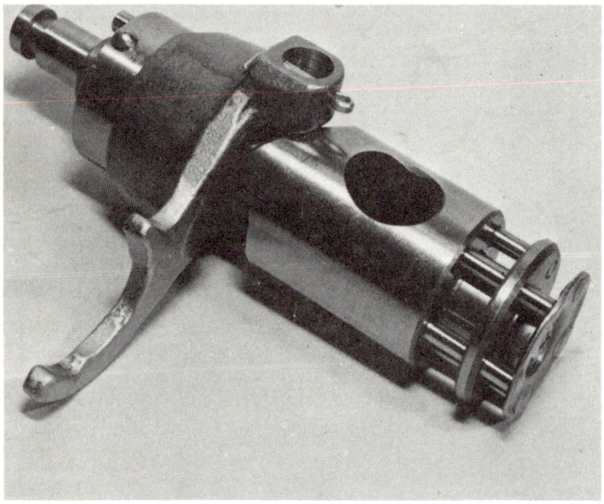
27.2e ... to allow selector fork to be removed

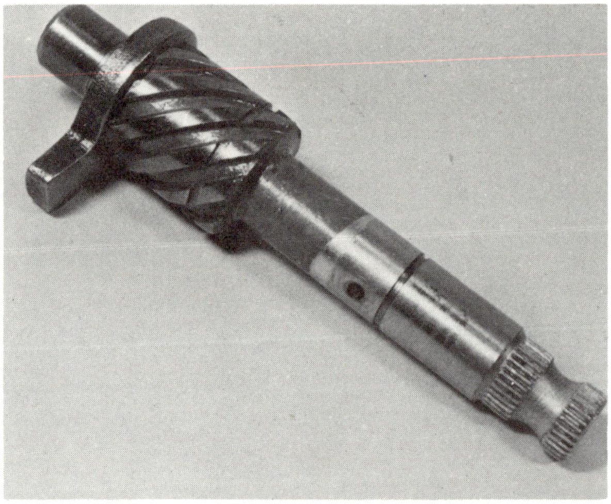
27.6a Check kickstart shaft for wear or damage ...

27.6b ... and check fit of pinion on thread

28 Clutch assembly: examination and renovation

1 After a considerable mileage has been covered, the bonded linings of the clutch friction plates will wear down to or beyond the specified wear limit, allowing the clutch to slip.
2 The degree of wear is measured across the faces of the friction material, the nominal, or new, size being 3.5 mm (0.14 in). If the plates have worn to 3.2 mm (0.13 in) they should be renewed, even if slipping is not yet apparent.
3 The plain plates should be free from scoring and signs of overheating, which will be apparent in the form of blueing. The plates should also be flat. If more than 0.05 mm (0.002 in) out of true, judder or snatch may result.
4 Measure the free length of the clutch springs which should be 28.2 mm (1.11 in) when new. Should the springs have compressed to 27.2 mm (1.07 in) or less they should be renewed.
5 Check the condition of the thrust bearing and washer and the bearing face on which it runs, on the clutch centre. Excessive play or wear will cause noise and erratic operation.
6 Check the condition of the slots in the outer surface of the clutch centre and the inner surfaces of the outer drum. In an extreme case, clutch chatter may have caused the tongues of the inserted plates to make indentations in the slots of the outer

drum, or the tongues of the plain plates to indent the slots of the clutch centre. These indentations will trap the clutch plates as they are freed, and impair clutch action. If the damage is only slight the indentations can be removed by careful work with a file and the burrs removed from the tongues of the clutch plates in a similar fashion. More extensive damage will necessitate renewal of the parts concerned.

7 The clutch release mechanism takes the form of a quick thread arrangement incorporated in the right-hand outer cover. No attention is normally required, and a considerable mileage can be covered before wear becomes apparent.

29 Gearbox components: reassembling the gear clusters

1 The mainshaft and layshaft gear clusters can be reassembled as described below. Reference should be made to the line drawing of the gearbox assembly and to the accompanying photographic sequence. Commencing with the layshaft fit the 4th gear pinion to the splined end. Working from the opposite end, fit the 2nd gear pinion, washer and circlip. The 3rd gear pinion should be fitted next, noting that the dogs should face the 2nd gear pinion. Finally, fit the large 1st gear pinion and its large shim washer. Note that the latter can be obtained in 0.6, 0.8 and 1.0 mm (0.024, 0.031 and 0.039 in) thickness, to provide end-float adjustment.

2 It will be noted that the gearbox mainshaft incorporates an integral 1st gear pinion. The remaining components are fitted from the left-hand (plain) end. Slide the 3rd gear pinion into position and secure it with its washer and circlip. Fit the 2nd gear pinion with the selector track inwards, followed by the large 4th gear pinion and its retaining circlip. The last two photographs in the sequence show the two clusters laid out in their respective positions, and the correct position of the selector drum and forks.

30 Engine and gearbox reassembly: general

1 Before reassembly of the engine/gear unit is commenced, the various component parts should be cleaned thoroughly and placed on a sheet of clean paper, close to the working area.

2 Make sure all traces of old gasket cement have been removed and that the mating surfaces are clean and undamaged. One of the best ways to remove some types of old gasket cement is to apply a rag soaked in methylated spirit. This acts as a solvent and will ensure that the cement is removed without resort to scraping and the consequent risk of damage. If scraping is necessary, use an aluminium or wooden scraper.

3 Gather together all the necessary tools and have available an oil can filled with clean engine oil. Make sure all the new gaskets and oil seals are to hand, also all replacement parts required. Nothing is more frustrating than having to stop in the middle of a reassembly sequence because a vital gasket or replacement has been overlooked.

4 Make sure that the reassembly area is clean and that there is adequate working space. Many of the smaller bolts are easily sheared if overtightened. Always use the correct size screwdriver bit for the crosshead screws and never an ordinary screwdriver or punch. If the existing screws show evidence of maltreatment in the past, it is advisable to renew them as a complete set.

5 If the purchase of a replacement set of screws is being contemplated, it is worthwhile considering a set of socket or Allen screws. These are invariably much more robust than the originals, and can be obtained in sets for most machines, in either black or nickel plated finishes. The manufacturers of these screw sets advertise regularly in the motorcycle press.

28.7 Clutch release rarely causes problems. Note ball staked into recess

29.1a Fit 4th gear pinion, then turn layshaft round ...

29.1b ... to fit 2nd gear pinion ...

29.1c ... which is retained by a washer and circlip

29.1d 3rd gear pinion is next to be fitted ...

29.1e ... followed by large 1st gear pinion ...

29.1f ... and its plain shim washer

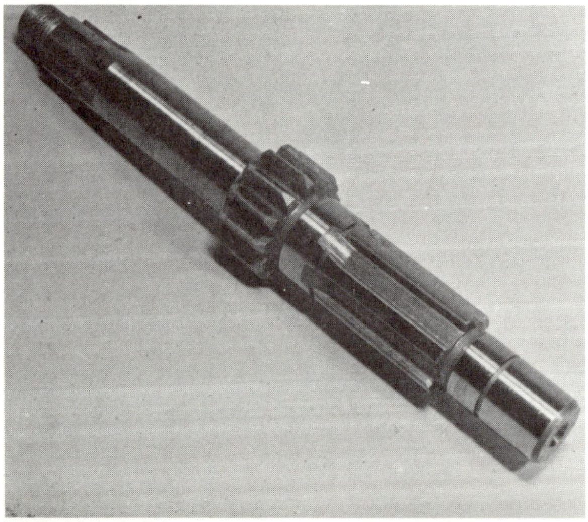

29.2a Mainshaft incorporates integral 1st gear pinion

29.2b Fit 3rd gear pinion (from plain end)

29.2c ... and secure with washer and circlip

29.2d Fit the 2nd gear pinion as shown ...

29.2e ... followed by 4th gear pinion

29.2f Do not omit to fit the circlip

29.2g Arrange assembled shafts as shown ...

29.2h ... then place selector drum and forks in position

31 Engine and gearbox reassembly: crankcase preparation

1 Check that the crankcase halves are perfectly clean. It is important that no residual dirt is allowed to remain inside the crankcase, as this will almost certainly cause significant damage. Check that the oil passage into the transfer port is clean and unobstructed.

2 Any new bearings should be fitted at this stage. It is recommended that the casings are heated to about 100°C (in an oven to avoid risk of warpage). Care must be taken to avoid burns when handling the heated casings. The bearings should be quite easy to fit. If necessary, use a suitably large socket as a drift, tapping the bearing gently into position, ensuring that the outer race locates squarely against the lip in the casing.

3 Do not forget to renew **all** of the oil seals on each casing after it has cooled down, if they have been heated for bearing renewal, as the heating will have damaged the existing seals, even if they have not been disturbed. When refitting the retaining plates, it is a good practice to apply a drop of locking fluid to the threads of each of the securing screws.

32 Engine and gearbox reassembly: joining the crankcase halves

1 Arrange the left-hand crankcase half on a workbench, placing wooden blocks beneath its edge to provide sufficient clearance for the crankshaft to protrude when fitted. Liberally oil the main bearing, then offer up the crankshaft assembly. Note that a shim washer is fitted at either end of the crankshaft to provide the correct end float clearance. These should be placed in position before the crankshaft is installed. Lower the crankshaft into position, ensuring that the connecting rod clears the crankcase mouth, and that the crankshaft seats squarely and fully into the main bearing.

2 Arrange the gearbox mainshaft and layshaft clusters, and the selector drum and forks, as an assembly, then lower them into the casing. Do not omit to place the shim over the right-hand end of the layshaft. Before proceeding further, thoroughly lubricate the gearbox components and bearings and the big-end bearing. Check that the two crankcase dowel pins are in position. These serve to locate the two crankcase halves accurately, and must not be omitted. It is important that an effective seal is made between the two halves of the crankcase to prevent oil leaks, and more important, to obviate any risk of a loss of crankcase pressure.

3 No crankcase gasket is used, so it is essential to use a good quality gasket cement, such as Golden Hermetite or Hylomar, or if available, Yamaha Bond No 4, to ensure a leak-free joint. An even film of the compound should be applied to the mating face, not forgetting the crank chamber (See Fig. 1.12).

4 Lubricate the bearings in the right-hand casing half, then offer it up to the assembled left-hand casing. If necessary, use a soft-faced mallet to tap the casing half into position. Note that the joint should close completely by hand – it should not be necessary to draw the two halves together with the casing screws. If the joint will not close, separate the two halves and establish why. Do not attempt to force the assembly if there is some reason for the joint not closing properly.

5 When the two casing halves are together, check that the crankshaft and gearbox shafts turn freely, then turn the unit over. Drop the casing screws into position, checking that they each protrude by approximately the same amount. As screws of various lengths are used, it may be necessary to interchange them until this is achieved. Tighten the screws evenly and in a diagonal sequence so that the casing halves are drawn together squarely. It is recommended that final tightening is undertaken using an impact driver. A certain amount of gasket cement will be squeezed out of the joint as the screws are tightened. This should be allowed to harden slightly, when it can be pared off with a knife. Make sure the crankshaft assembly revolves quite freely.

33 Engine and gearbox reassembly: fitting the disc valve

1 Turn the crankshaft until it is at the top dead centre (TDC) position, at which point the connecting rod will be centrally disposed in the crankcase mouth. The small driving pin should now be fitted so that it faces the crankcase mouth and is in line with the connecting rod. It is important that the pin is fitted correctly, otherwise the valve disc will be 180° out, and the reassembled engine will not run.

2 As mentioned in Section 15, there are two basic types of valve fitting, and the valve in question should be positioned

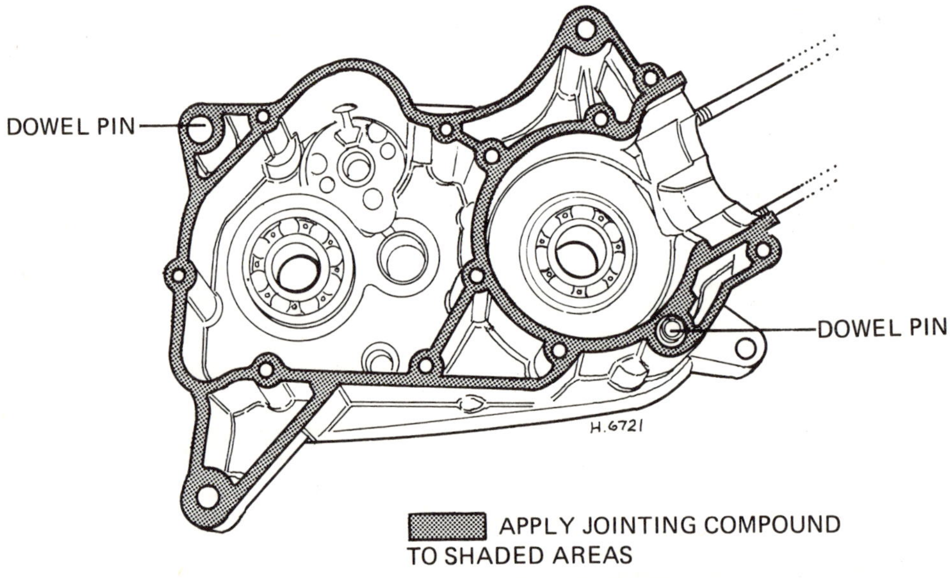

Fig. 1.13 Jointing compound application

noting the remarks made during removal. Ensure that the O-ring on the crankshaft is in good condition, and renew it if marked or cut in any way. Wipe the O-ring with grease before fitting the valve disc assembly.
3 When the valve disc is fitted, slowly rotate the crankshaft and check that the inlet port is uncovered as the crankshaft passes TDC. If this is not the case, check that the driving pin is correctly positioned, and on the later, splined fitting, that the index marks are aligned.
4 If all is well, liberally oil the valve disc. Check that the two locating dowel pins are in position. Examine the large O-ring seal in the valve cover. If this has been removed from its groove, but is in perfect condition and is to be reused, it can be refitted as follows. Lay the O-ring on its groove, then press it into position at about four points around the groove. Work the rest of the O-ring into its groove gradually. If the O-ring is fitted by starting at one point it is likely that there will be rather too much O-ring to fit in the groove.
5 Fit the cover carefully, having first greased the lip of the crankshaft seal. Fit the retaining screws, and tighten them down in a diagonal sequence to avoid warpage.

32.1a Note O-ring and thrust washers on crankshaft

32.1b Install the crankshaft in left-hand casing ...

32.1c ... ensuring that thrust washers are in place

32.2a Gearbox cluster is installed as shown ...

32.2b ... not forgetting the large plain washer

32.4 Apply jointing compound, fit dowels, and offer up right-hand casing

33.1 Pin (arrowed) must align with connecting rod at TDC

33.2a Renew O-ring unless unmarked

33.2b Ensure that valve disc is fitted correctly ...

33.5 ... then install cover

34 Engine and gearbox reassembly: refitting the selector mechanism

1 If the gearbox mainshaft bearing retainer has been removed, this should be refitted, and the securing screws tightened. Fit the neutral stopper arm, retaining it with its shouldered washer. Engage the stopper spring in the end of the arm, then hook the free end to the hole in the mainshaft bearing retainer.

2 Select neutral by turning the end of the selector drum fully clockwise, whilst turning the gearbox mainshaft to facilitate selection. Turn the drum back by one stop. This should be the neutral position which can be checked by holding the layshaft end and turning the mainshaft. There will be a tendency for both to turn, due to drag between the gear pinions and shafts, but there should be no direct drive present.

3 Whilst the unit is upright to permit the above check, screw the neutral switch body into the casing. Fit the C-shaped retainer plate to the end of the selector drum, then fit the white plastic cover which houses it, noting that the rearmost of the three securing screws also retains a clip for the generator output leads.

4 Slide the selector claw and cross-over shaft assembly into place, lifting the claw into engagement with the pins on the end of the selector drum and fitting the ends of the centring spring over its pin. Fit the circlip to the left-hand end of the cross-over shaft, and temporarily refit the gearchange pedal. Check that all four gears select properly. It will be necessary to turn the gearbox mainshaft end to ease selection. Find 1st, 2nd or 3rd gear, then check the selector claw adjustment.

5 Examine the position of the two jaws of the selector claw in relation to the pins in the end of the selector drum. Each jaw should be a similar distance from the nearest pin. If this is not the case, slacken the locknut on the centring spring locating pin, and turn the pin until the correct setting is obtained. Do not omit to secure the locknut after this adjustment has been made.

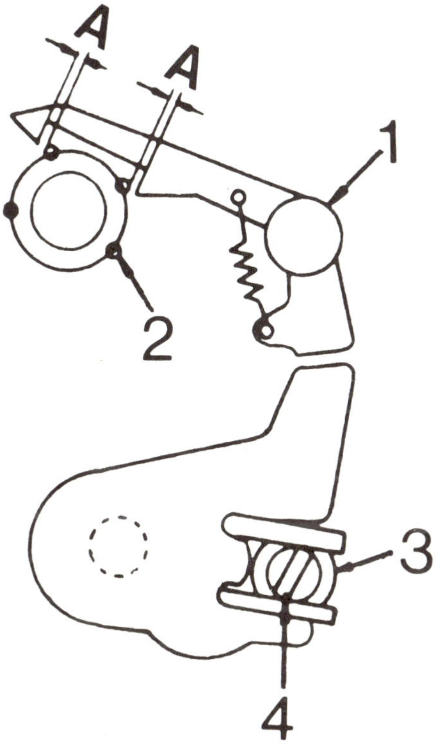

Fig. 1.14 Setting the selector arm

A and A should be equidistant
1 Selector arm
2 Selector drum pins
3 Adjuster locknut
4 Eccentric adjuster

35 Engine and gearbox reassembly: refitting the kickstart mechanism

1 Fit the plain washer to the protruding end of the gearbox layshaft, followed by the thin wave washer. Install the kickstart idler pinion and fit its retaining circlip. If the kickstart shaft components have been removed for examination or renewal, these should be reassembled in the reverse order of that described in Section 13.

2 Install the assembled shaft into the casing, ensuring that the friction clip locates in the recess in the casing. Grasp the end of the return spring, and twist it until it can be engaged with its anchor pin. Lubricate the moving parts of the kickstart mechanism with clean engine oil.

34.1a Refit bearing retainer plate ...

34.1b ... and fit stopper mechanism as shown

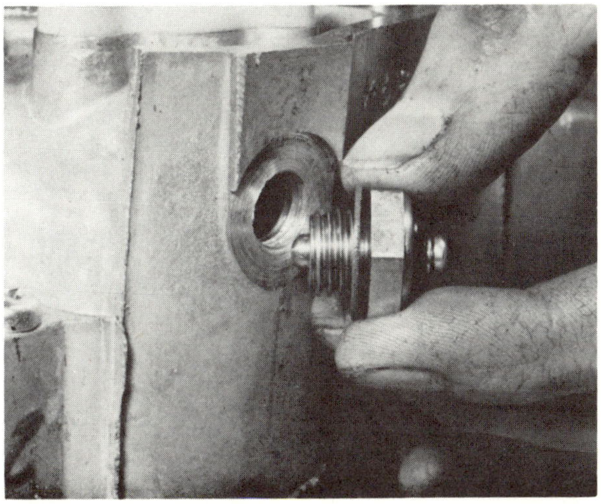

34.3a Fit neutral light switch

34.3b Selector drum is retained by C-collar and cover

34.4a Slide selector shaft through bore

34.4b Centring spring and claw must fit as shown

34.4c Do not forget clip at left-hand end

35.1a Fit plain and wave washers, ...

35.1b ... followed by idler gear and circlip

35.1c Reassemble kickstart mechanism and install, ...

35.2 ... making sure that friction clip enters recess

36 Engine and gearbox reassembly: refitting the clutch and crankshaft pinion

1 On early models, fit the plain washer and bearing sleeve, lubricating the latter with engine oil. Slide the kickstart driven pinion into position, noting that the dogs face outwards. Later models employ a plain washer and a short bearing sleeve, the driven pinion being integral with the clutch drum. On either type, slide the clutch drum into place.

2 Fit the thrust bearing into position. Note that the early type comprises a needle roller race sandwiched between two large plain washers, whilst the later type consists of a single plain thrust washer. Fit the inner pressure plate. Fit the clutch plates, starting with a friction plate, then a plain plate, building the assembly up in layers. On early models, rubber rings are fitted between the plain plates to aid freeing when the clutch is disengaged. These are not used on later clutches.

3 On early models, the single, thick, plain plate should be fitted last, followed by the clutch centre. On later machines, the two components are combined in a single unit. It will be found in practice that it is convenient to assemble the inner pressure plate, the clutch plates and the clutch centre. The assembled unit can now be installed in the clutch drum. Note that on the late type clutch, it is important to align the index marks on the inner pressure plate and the clutch centre.

4 Fit the Belville washer with its cupped face towards the clutch centre, then fit and tighten the securing nut, employing one of the methods described in Section 12 to prevent the clutch centre from turning. The securing nut should be tightened to 4.5 kgf m (32 lbf ft).

5 Fit the six clutch springs, followed by the pressure plate. The plate is retained by six cross-headed screws. These should be fitted and tightened in a diagonal sequence to ensure that the pressure plate is pulled down evenly.

6 With the clutch assembly completed, fit the crankshaft pinion into place, followed by the plain washer (keyed pinion) or spring washer (splined pinion). Immobilise the crankshaft by one of the methods described earlier in this Chapter, and tighten the retaining nut to 6 kgf m (43 lbf ft).

36.1a Fit washer and clutch sleeve

36.1b Early models use separate pinion ...

36.1c ... which engages dogs at rear of clutch drum

36.1d Fit clutch drum over shaft end

36.2 Early models have needle roller thrust race

36.3a Early machines have rubber rings between plates

36.3b Assemble clutch centre and plates, then install

36.4a Lock clutch as shown, fit washer and nut ...

36.4b ... and tighten nut securely

36.5a Fit the six clutch springs and cover

36.5b Tighten screws down evenly

36.6 Fit and secure crankshaft pinion (Clutch not shown)

37 Engine and gearbox reassembly: refitting the right-hand inner cover

1 Check that the kickstart mechanism, disc valve assembly, clutch and gear selector mechanism are all correctly installed. If the clutch release mechanism has been removed from the cover, this should be refitted. Carefully clean the mating faces of the cover and crankcase, removing any residual jointing compound or gasket. Fit the locating dowel pins into position.

2 Apply a light coating of good quality non-hardening gasket cement, such as Golden Hermetite, to stick the new cover gasket into position. Wrap one or two turns of PVC tape around the kickstart shaft splines to avoid damage to the seal during installation. Fit a new O-ring to each of the two projections on the disc valve cover. Offer up the casing, checking that it seats squarely around the gasket face. Refit the securing screws, tightening them in a diagonal sequence to avoid warpage.

37.2 Fit new O-rings to disc valve stubs – grease helps seal

38 Engine and gearbox reassembly: refitting the flywheel generator

1 If the flywheel generator was removed from the engine unit prior to its removal from the frame, this operation may be postponed until the unit has been installed. Offer up the generator stator, and fit the securing screws. The generator output leads should be routed through the guide slots in the crankcase, ensuring that the grommet is located properly and that the leads pass through the guide clip attached to the rear most of the three screws which retain the white plastic cover at the end of the selector drum.
2 It will be noted that there is no provision for altering the ignition timing by moving the generator stator. This does not mean that the timing can be assumed to be correct, and this should be checked before the machine is used. See Chapter 3 for details.
3 Fit the Woodruff key into its slot in the crankshaft end, then offer up the generator rotor, checking that it locates accurately on its key. Fit the washer and retaining nut, and hold the crankshaft to prevent rotation, using one of the methods described in Section 5. Tighten the nut to 5·0 kgf m (36 lbf ft).

39 Piston, cylinder barrel and cylinder head: reassembly

1 Ensure the mating faces of all components are clean and dry.
2 Fit a new cylinder base gasket and lubricate the big end bearing with a little two-stroke oil. Do NOT use gasket cement at this joint.
3 Lubricate the small end needle bearing and fit to the connecting rod.
4 Fit the piston and rings, with the arrow on the piston crown facing towards the exhaust port. Fit the gudgeon pin and secure it with the two new circlips, making sure they have seated properly in their grooves. It is bad policy to use old circlips as they have lost much of their spring tension and may jump out and cause the engine to seize.
5 Remove any rag from around the connecting rod, lubricate the piston, rings and bore of the cylinder barrel. Align the piston ring gaps over their locating pegs and slide the cylinder barrel over the piston and down over the holding down studs, whilst compressing the piston rings with the fingers to assist assembly.
6 When the cylinder barrel is seated correctly, fit a new

38.1 Note cable clip on cover screw

38.3 Rotor is located by a Woodruff key

39.3 Lubricate and fit small end bearings

39.4a Fit the piston ...

39.4b ... noting arrow which must face exhaust port

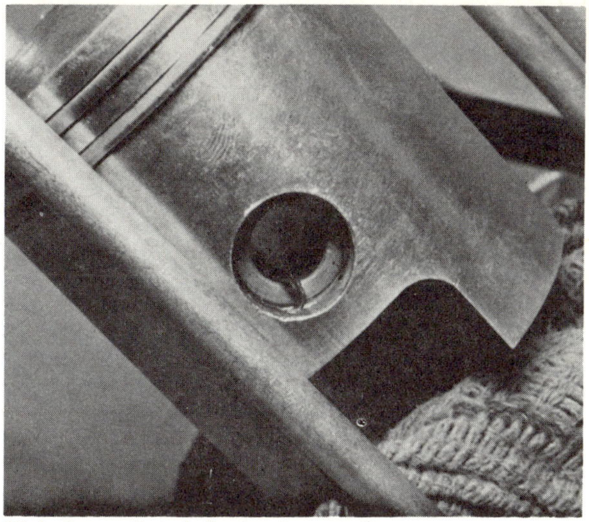
39.4c Ensure that circlips locate correctly

39.5a Check that the rings are correctly positioned

39.5b Feed piston rings into bore

39.6 Fit a new cylinder head gasket ...

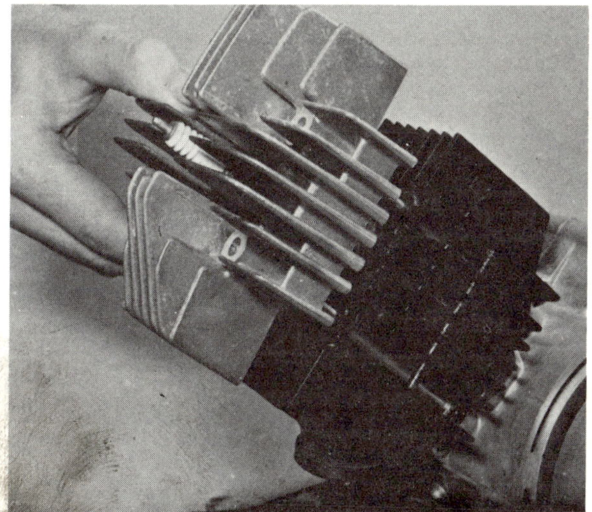
39.7 ... then offer up the cylinder head

cylinder head gasket. Do NOT use any gasket cement at this joint.
7 Fit the cylinder head, with the large chamfered edge facing downwards. Secure the cylinder head with the four washers and nuts. Torque tighten the nuts to 1·4 kgf m (10 lbf ft) in a diagonal sequence. Squirt a little two-stroke oil into the plug hole and turn the engine over a few times. Replace the sparking plug.

40 Replacing the engine unit in the frame

1 Installation of the rebuilt engine/gearbox unit is, generally speaking, a straight-forward reversal of the removal sequence. Where special attention should be paid to a particular assembly sequence, this is given in the following paragraphs.
2 Lift the engine unit into position, and slide the lower rear mounting bolt into place to support the rear of the unit. Raise the front of the unit slightly by placing some wooden blocks below the front of the crankcase. If it has not been fitted at this stage, replace the flywheel generator assembly, following the instructions given in Section 38 of this Chapter. The neutral switch lead must be fitted at this stage, as it is impossible to do so after the engine is properly installed. This situation also exists with the generator output leads on early models, as once the engine unit is in place it is almost impossible to feed the leads up into the frame. Thread the leads up through the open end of the frame section until they can be reached via the battery aperture. Reconnect the leads to the main wiring harness, paying attention to the colour coding of the wires.
3 Lift the front of the engine unit and fit the two upper engine mounting bolts. Withdraw the lower mounting bolt whilst the footrest and propstand assembly is repositioned, then refit it and the single locating bolt on the left-hand side of the machine. Note that the air filter casing is secured by the front engine mounting bolt, and must therefore be positioned before the bolt is tightened. When the air cleaner is in position, fit the remaining bolt, and tighten this and the engine mounting bolts.
4 The final drive chain can be installed next, together with the gearbox sprocket, if this has been removed. When fitting the sprocket, check first that the spacer has been positioned. The oil seal in which it runs should be greased before it is inserted. If the chain has been removed for cleaning, run it on to the sprocket at the rear wheel. Fit the gearbox sprocket, tab washer and securing nut. The chain joining link can now be assembled, noting that the closed end of the spring clip must face in the same direction as normal chain travel. The rear brake can be applied to prevent the sprocket from turning whilst the securing nut is tightened. Do not omit to knock over the locking tab. Do not fit the outer cover until the contact breaker gap/ignition timing setting has been checked as described in Chapter 3, Section 2.
5 The pressed steel (plastic on later models) battery holding bracket can be refitted in the recess in the left-hand side of the frame. Refit the battery but do not connect it up as a precaution against accidental short circuits. On later machines, plug in the generator connector. If the final drive chain enclosure was removed, this can now be refitted. The exhaust pipe may be fitted next, preferably with a new gasket at each end. Both ends of the pipe are secured by large sleeve nuts, and these should be tightened with the C-spanner supplied with the machine. In its absence, a large worm drive hose clip can be tightened around the nut, and this can be used to obtain a good grip with a pair of Stilsons or a Mole wrench.
6 Feed the clutch operating cable into the casing, and assemble the operating arm over the nipple. Fit the arm over the projecting flats of the release mechanism, and retain it with the adjuster locknut. Before securing the locknut, set the clutch adjustment by slackening the adjuster locknut by about one turn, and holding this position with a spanner. Slowly screw the adjuster inwards until some resistance is felt, indicating that all free play has been taken up. If necessary, experiment a little to get the feel of this point. The screw should be backed off by $\frac{1}{4}$ turn from the above point, and then held in this position while the locknut is retightened. Moving to the cable adjuster on the outside of the casing, slacken the locknut and set the adjuster to give 2–3 mm (0·08 – 0·12 in) free play measured between the handlebar lever stock and blade.
7 Reconnect the oil feed pipe to the oil tank outlet, and the delivery pipe at the casing union. Use new sealing washers to ensure a leak-proof joint. Insert the oil pump cable nipple into the hole in the pulley on the oil pump, and screw the cable adjuster into position. Do not tighten the adjuster locknut at this stage, as it will be necessary to synchronise the oil pump with the carburettor later on. Ensure that the oil pump cable grommet fits squarely into its casing recess.
8 The carburettor can now be installed in its recess, noting that it will be necessary to refit the breather, overflow and feed pipes as it is installed. Fit the carburettor over the projecting disc valve stub, and secure it by passing a screwdriver through the access hole at the front of the casing. Tighten the clamp screw, then refit the small rubber blanking plug. Fit the air cleaner hoses between the crankcase and air cleaner case, and between the air cleaner case and the frame.
9 The throttle and oil pump cables should be adjusted together, in the sequence described below, to ensure that the oil pump is accurately synchronised to the throttle valve opening. The throttle cable is in two sections. The upper section runs down from the twistgrip control to a cable splitter. From this unit, the lower throttle cable and the oil pump control cables emerge. Set the lower throttle cable free play first, to give 1·0 mm (0·04 in) movement. Next, note the amount of movement, measured at the flange of the throttle twistgrip, before the cable slack is taken up. This should be 3 – 6 mm (0·12 – 0·24 in) and can be altered by using the adjuster at the top of the upper cable.
10 Having set the throttle cable free play, attention may be turned to the oil pump cable setting. It will be seen that the throttle valve (viewed through the mouth of the carburettor) has a circular mark on it. The throttle should be opened until the upper edge of the circle touches the edge of the main bore (see accompanying illustration). At this point, the raised index mark on the pump pulley must align with the plunger pin. If necessary, adjust the cable until this setting is obtained, then tighten the locknut. Open and close the throttle a few times, and recheck the setting.
11 Check that the oil tank contains plenty of oil, then bleed the air from the Autolube system. The bleeding operation is undertaken in two stages, the first being to bleed the pipe from the oil tank to the pump and the pump body. Have a small bowl or some rag to catch surplus oil. Remove the bleed screw from the pump body. This screw can be identified by its sealing washer. Allow the air to be forced out by the oil running down from the oil tank, until the emerging oil is free of air bubbles. When all the air has been cleared, refit and tighten the bleed screw.
12 The second stage of the bleeding operation is to clear the pump distributor and the delivery pipe, and must be done with the engine running. The procedure is quite simple, and involves pulling the pump cable to set the pump at full delivery capacity. The engine should be allowed to run for about two minutes at a fast idle to clear any air bubbles in the delivery side of the pump. This operation can be undertaken after the initial start up has been made, and the rebuilt engine has been checked for leakages.
13 Remove the gearbox filler plug, and fill to within the level marks on the dipstick using SAE 10W/30 engine oil. Note that the level is measured with the filler plug resting in position, and **not** screwed fully home. The gearbox and transmission casing will hold about 700 cc (1·23 Imp pint, 1·48 US pint) when filling a newly rebuilt unit, and subsequent oil changes will require about 600 cc (1·06 Imp pint, 1·27 US pint). Check the oil level again after running the engine for the first time.
14 Finally, refit the sparking plug lead and reconnect the battery. Check that the electrical system functions correctly before any attempt is made to start the engine. Do not forget to

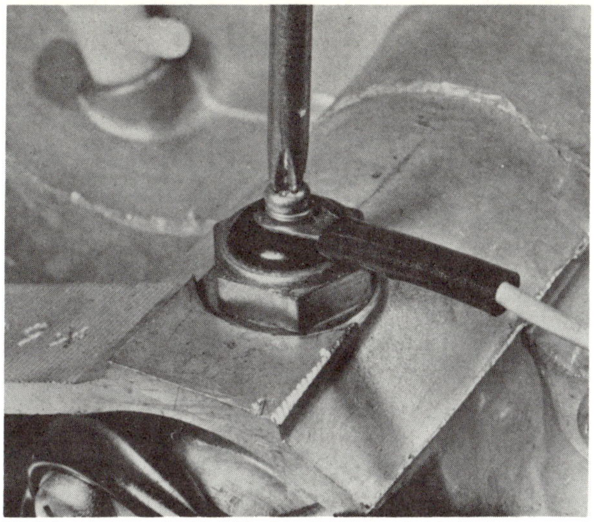
40.2 Connect neutral lead before fitting engine bolts

40.3 Engine rear mounting bolts (arrowed)

40.4a Do not omit spacer when fitting gearbox sprocket

40.4b Recess on nut must face inwards

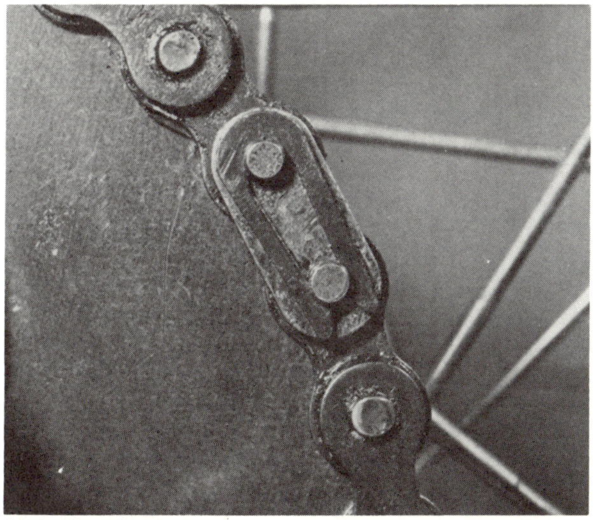
40.4c Closed end of clip must face direction of chain travel

40.5a Reassemble the chain enclosure halves

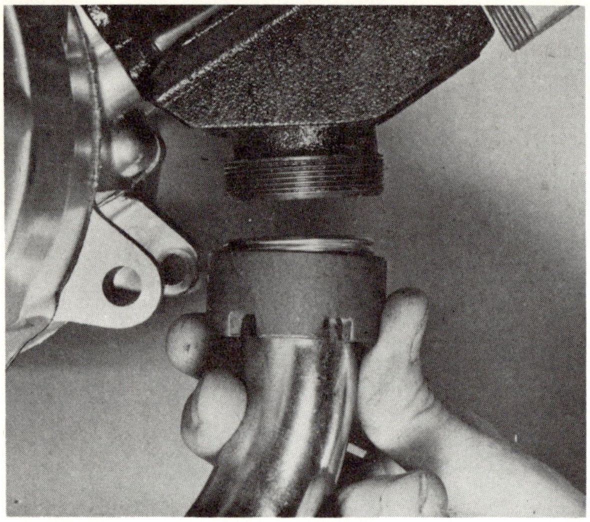
40.5b Exhaust pipe is retained by sleeve nuts ...

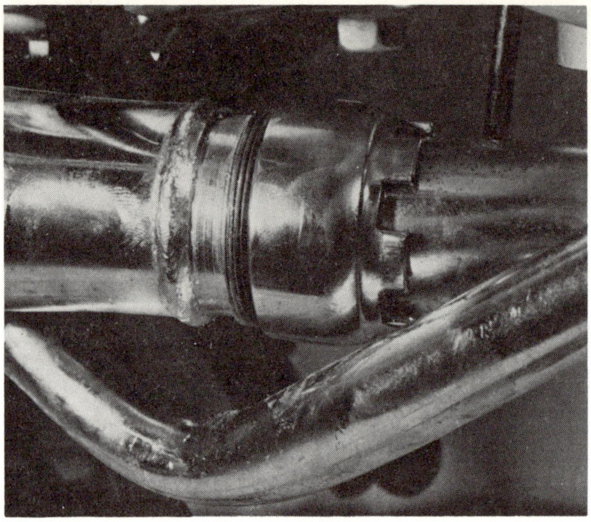
40.5c ... at each end. Use new gaskets

40.6a Adjust clutch mechanism ...

40.6b ... and cable free play

40.13 Top up gearbox to recommended level

check the ignition timing/contact breaker gap setting, referring to Chapter 3 for details. If the petrol tank and seat were removed, these can be refitted and the petrol feed and balance pipes reconnected.

41 Starting and running the rebuilt engine

1 When the initial start-up is made, run the engine slowly for the first few minutes, especially if the engine has been rebored or a new crankshaft fitted. Check that all the controls function correctly and that there are no oil leaks before taking the machine on the road. The exhaust will emit a high proportion of white smoke during the first few miles, as the excess oil used whilst the engine was reassembled is burnt away. The volume of smoke should gradually diminish until only the customary light blue haze is observed during normal running. It is wise to carry a spare sparking plug during the first run, since the existing plug may oil up due to the temporary excess of oil.

2 Remember that a good seal between the piston and the cylinder barrel is essential for the correct functioning of the engine. A rebored two-stroke engine will require more careful running-in, over a long period, than its four-stroke counterpart. There is a far greater risk of engine seizure during the first hundred miles if the engine is permitted to work hard.

3 Do not tamper with the exhaust system or run the engine without the baffle fitted to the silencer. Unwarranted changes in the exhaust system will have a marked effect on engine performance, invariably for the worse. The same advice applies to dispensing with the air cleaner or the air cleaner element.

4 Do not on any account add oil to the petrol under the mistaken belief that a little extra oil will improve the engine lubrication. Apart from creating excess smoke, the addition of oil will make the mixture much weaker, with the consequent risk of overheating and engine seizure. The oil pump alone should provide full engine lubrication.

42 Fault diagnosis: engine

Symptom	Cause	Remedy
Engine will not start	Defective sparking plug	Remove plug and lay it on cylinder head. Check whether spark occurs when engine is kicked over.
	Dirty or closed contact breaker points	Check condition of points and whether gap is correct.
	Air leak at crankcase or worn crankshaft oil seals	Check whether petrol is reaching the sparking plug.
	Air leak between disc valve cover and inner cover	Check condition of O-ring and renew if necessary.
Engine runs unevenly	Ignition and/or fuel system fault	Check systems independently as though engine will not start.
	Blowing cylinder head gasket	Leak should be evident from oil leakage where gas escapes.
	Incorrect ignition timing	Check timing very accurately and reset if necessary.
	Worn crankshaft seals or O-rings	See above
Lack of power	Fault in fuel system or incorrect ignition timing	See above.
	Choked silencer	Remove and clean out baffles.
High fuel/oil consumption	Cylinder barrel in need of rebore and o/s piston	Fit new rings and piston after rebore.
	Oil leaks or air leaks from damaged gaskets or oil seals	Trace source of leak and replace damaged gasket and/or seal.
	Oil pump stroke adjustment incorrect	See Chapter 2 Section 11.
	Oil pump cable setting incorrect	See Chapter 2 Section 11.
Excessive mechanical noise	Worn cylinder barrel (piston slap)	Rebore and fit o/s piston.
	Worn small end bearings (rattle)	Replace needle roller bearing (caged) and if necessary, gudgeon pin
	Worn big end bearings (knock)	Fit replacement crankshaft assembly.
	Worn main bearings (rumble)	Fit new journal bearings and seals.
Engine overheats and fades	Pre-ignition and/or weak mixture	Check carburettor settings. Check also whether plug grade is correct.
	Lubrication failure	Check oil pump setting and whether oil tank is empty.

Fault diagnosis: Clutch and Gearbox overleaf

43 Fault diagnosis: clutch

Symptom	Cause	Remedy
Engine speed increases but machine does not respond	Clutch slip	Check clutch adjustment for free play at handlebar lever. Check condition of clutch plate linings.
Difficulty in engaging gears. Gear changes jerky and machine creeps forward, even when clutch is withdrawn. Difficulty in selecting neutral	Clutch drag	Check clutch adjustment for too much free play. Check for burrs on clutch plate tongues or indentations in clutch drum slots. Dress with file if damage not too great.
	Clutch assembly loose on mainshaft	Check tightness of retaining nut. If loose, fit new tab washer and retighten.
Operating action stiff	Damaged, trapped or frayed control cable	Check cable and replace if necessary. Make sure cable is lubricated and has no sharp bends.

44 Fault diagnosis: gearbox

Symptom	Cause	Remedy
Difficulty in engaging gears	Gear selector forks bent	Renew.
	Gear cluster assembled incorrectly	Check that thrust washers are located correctly.
Machine jumps out of gear	Worn dogs on ends of gear pinions	Renew pinions involved.
	Selector drum pawls stuck	Free pawl assembly.
Gear lever does not return to normal position	Broken return spring	Renew spring.
Kickstart does not return when engine is turned over or started	Broken or poorly tensioned return spring	Renew spring or retension.
Kickstart slips	Kickstart drive pinion internals, pawls or springs worn badly	Renew all worn parts.

Chapter 2 Fuel system and lubrication

Contents

General description	1
Petrol tank: removal and replacement	2
Fuel tap	3
Carburettor: removal and replacement	4
Carburettor: dismantling, examination and renovation	5
Carburettor: reassembly	6
Carburettor: checking the settings	7
Carburettor: adjustment	8
Disc valve induction system: examination and renovation	9
Air filter: removal and cleaning	10
Engine lubrication system: maintenance and adjustment	11
Oil pump: removal and replacement	12
Bleeding the oil pump	13
Silencer: decarbonisation	14
Fault diagnosis: fuel system	15
Fault diagnosis: lubrication system	16

Specifications

	YB100	YB100E
Carburettor		
Make	Mikuni	Mikuni
Type	VM20SC	VM20SC
Main jet	115	125
Pilot jet	25	22.5
Starter jet	30	30
Jet needle	4J26	4L5
Needle position	3rd groove	2nd groove
Needle jet	N–6	N–4
Throttle valve	2.5	2.5
Pilot screw setting	$1\frac{1}{2}$ turns out	$1\frac{1}{4}$ turns out
Float height	21.0 ± 2.5 mm (0.83 ± 0.1 in)	
Idle speed	1200 – 1300 rpm	
Engine lubrication		
System	Yamaha Autolube	
Pump colour code	Yellow	
Minimum stroke	0.20–0.25 mm (0.0078–0.0098 in)	
Maximum stroke	1.85–2.05 mm (0.008–0.010 in)	
Minimum output	0.50–0.63 cc per 200 strokes	
Maximum output	4.65–5.15 cc per 200 strokes	
Oil tank capacity	1.2 litre (2.11 Imp pint/2.54 US pint)	
Oil grade	2-stroke engine oil	

1 General description

Petrol is gravity-fed from the petrol tank to the float chamber of the Mikuni carburettor via a three position fuel tap which incorporates a small filter. Air is drawn in through an oil impregnated filter element housed in a cylindrical canister, mounted above the crankcase.

At low engine speeds, the proportions of air and atomised petrol which form the combustion mixture, are controlled by a pilot circuit, these being regulated by a combination of throttle stop and pilot mixture screw settings. As the twistgrip control is turned, the cylindrical throttle valve is lifted, allowing a greater volume of air to be drawn through the carburettor choke. The passage of air across the top of the needle jet causes fuel to be drawn up through the main jet and needle jet by venturi action.

The amount of fuel entering the engine is at this stage metered by the needle jet assembly, in which a tapered needle is drawn upwards with the throttle valve, allowing increasing amounts of fuel to enter the combustion mixture, as the throttle is opened. Eventually, the rate of flow of the fuel is restricted by the main jet, which has been selected to give the correct mixture at maximum throttle opening.

The point of induction on a two-stroke engine is normally controlled by the piston skirt, which covers and uncovers ports machined in the cylinder bore. On Yamaha YB 100 models a supplementary timing system is employed to enable more efficient induction timing. This device is known as a disc valve, and consists of a thin fibre disc, which is mounted on the crankshaft and enclosed by an alloy casing.

The casing has a port machined in it, and this aligns with a similar cutout in the valve disc. At a predetermined crankshaft position, the valve opens, allowing the combustion mixture to be drawn into the crankcase in the usual way. The valve then closes, preventing any back leakage of the mixture. The mixture is then fed to the combustion chamber via transfer ports in the normal manner.

Engine lubrication is by a pump fed system known as Yamaha Autolube. Two-stroke oil is gravity fed from a frame-mounted tank to the oil pump. The pump is driven by the engine via the primary and kickstart drive assemblies, and is also interconnected by a Bowden cable to the throttle twist grip. Thus the amount of oil passed by the pump is varied according to the engine speed and throttle setting.

2 Petrol tank: removal and replacement

1 Various petrol tanks have been used on the YB 100 models, depending upon the year of manufacture and the country to which the machine was exported. The tank is retained at the front by two rubber buffers, which engage in 'C'-shaped brackets below the earlier type of tank, and in cups inside the saddle of the later type. The rear of the tank is secured by two lugs retained by a nut and bolt, access to which may necessitate the removal of the dualseat, depending on the model.

2 Before removing the tank, ensure that the petrol tap is in the 'off' position, and that the petrol feed pipe is disconnected. It will also be necessary to remove the small bridge pipe which joins the tank halves. Obviously, this will necessitate draining the content of the tank. Whilst the tank is removed, check for signs of leakage or damage. When refitting the tank, check that no electrical or control cables are trapped beneath the tank saddle.

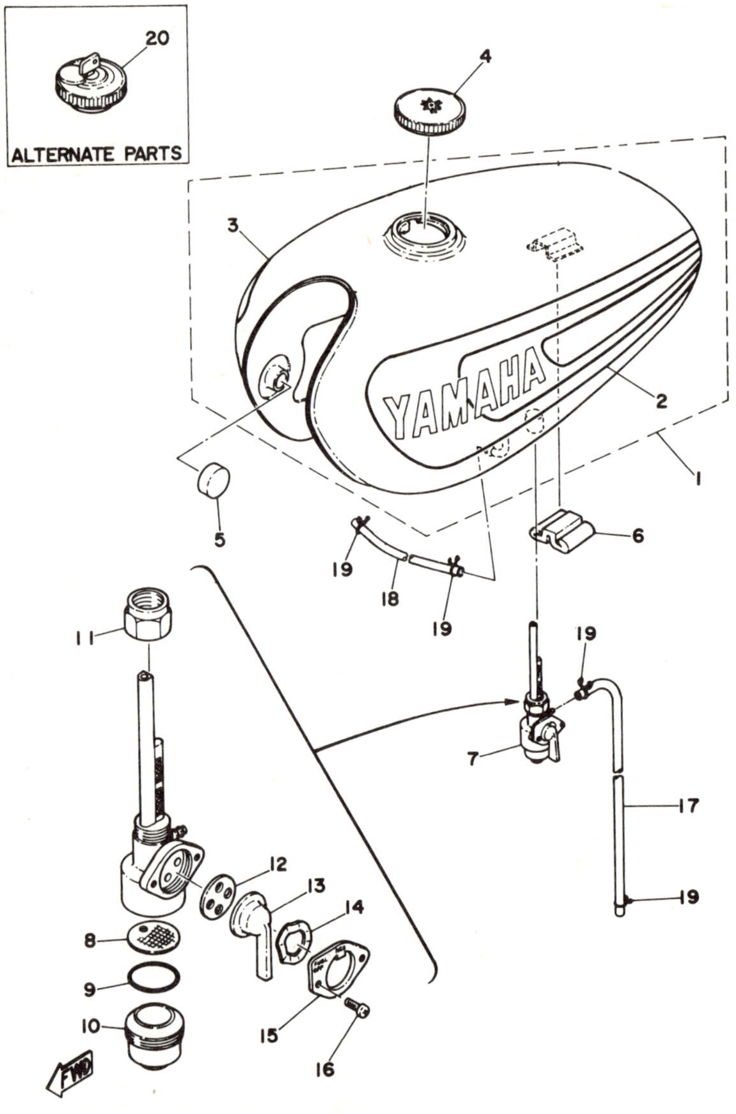

Fig. 2.1 Fuel tank and tap

1 Fuel tank
2 Decal (left-hand)
3 Decal (right-hand)
4 Filler cap
5 Locating rubber – 4 off
6 Locating rubber
7 Complete fuel tap
8 Filter gauze
9 Sealing washer
10 Filter bowl
11 Nut
12 Valve
13 Fuel tap lever
14 Wave washer
15 Fuel tap plate
16 Screw – 2 off
17 Fuel pipe
18 Level pipe
19 Pipe clip – 4 off
20 Alternative petrol cap

3 Fuel tap

1 The sediment bowl of the fuel tap should be cleaned out periodically. It is removed by unscrewing. When refitting, check that the upper securing bolts are not loose and leaking. Tighten as necessary. Also, check the 'O' ring seal.
2 If the fuel tap is leaking, it is most likely due to damage or deterioration of the rubber seals. To gain access to them remove the two screws from the lever retaining plate. Lift off the plate and 'O'-ring and pull out the tap lever. If necessary, renew the rubber seals. The tank must be drained of fuel for this operation.
3 The fuel tap is removed by slackening the gland nut and carefully guiding the tap stem from out of the tank. The tank must be drained for this operation. Replacement is the reverse of removal.

4 Carburettor: removal and replacement

1 As a result of the incorporation of a disc valve assembly, the incoming mixture is drawn in via a tract in the right-hand crankcase. Consequently the carburettor is contained within a compartment at the front of the right-hand outer casing.
2 Remove the inspection cover, which is retained by four cross-headed screws. The rubber cover and its retaining rim, at the top of the chamber should also be released, and slid up the operating cables to permit the carburettor to be partially withdrawn. Prise off the petrol feed pipe (turn the petrol tap off first) and also the two small vent tubes.
3 Pull out the blanking plug at the front of the casing. This allows access to the carburettor clamping screw, which should be slackened, using a screwdriver. Pull the carburettor off the mounting stub, and partially withdraw it from the chamber. This will permit the mixing chamber top to be released by unscrewing the retaining ring. Unless specific attention is required, the throttle valve assembly can be pulled out of the carburettor and allowed to hang on the operating cables. Unscrew the starting device, which can also be left to hang from its cable. The main body of the instrument may now be removed for examination.
4 Replacement is a straightforward reversal of the removal operation, ensuring that the carburettor is mounted vertically in the chamber. Check the carburettor settings before refitting the inspection cover.

3.1a Filter is mounted in tap body (tap removed for clarity)

3.1b Gauze filter can be removed for cleaning

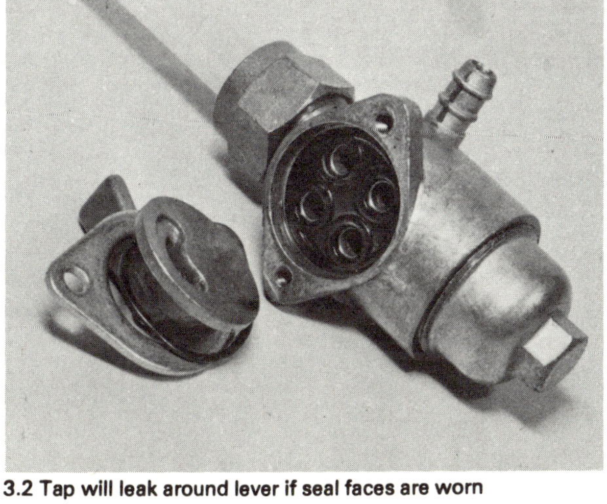
3.2 Tap will leak around lever if seal faces are worn

3.3a Drain petrol tank before removing tap

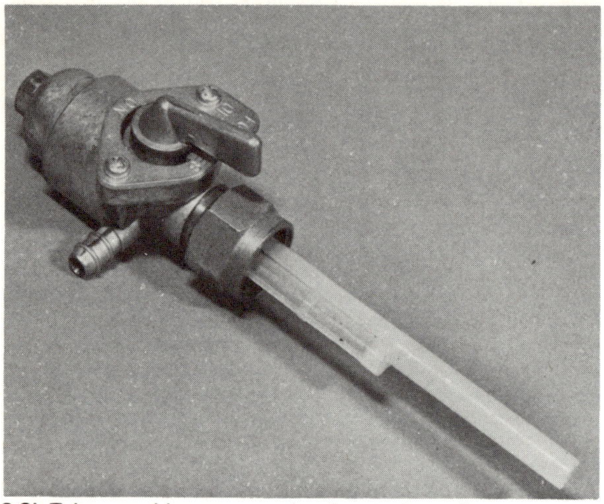
3.3b Tubes provide reserve and main petrol supplies

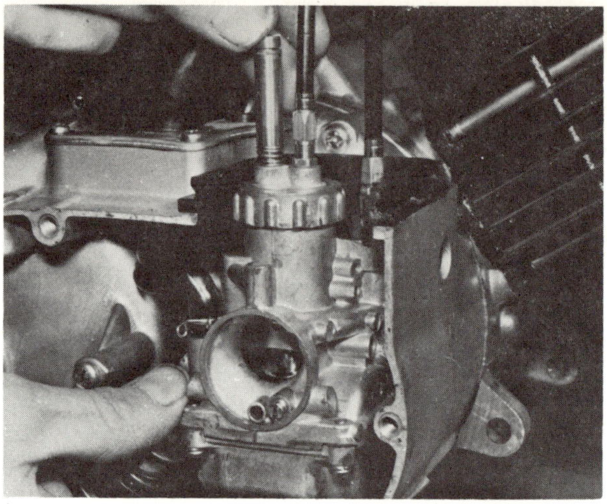
4.3a Withdraw carburettor and release the various pipes

4.3b Carburettor top can be unscrewed and valve withdrawn

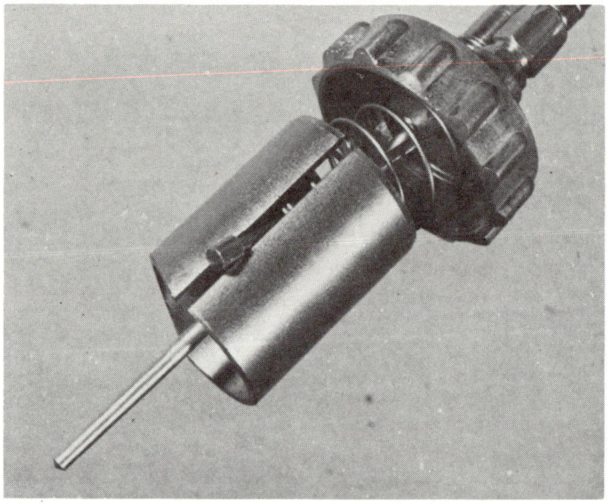
4.3c Valve may be released by displacing cable as shown

5 Carburettor: dismantling, examination and renovation

1 Remove the four screws and lift off the float bowl which will uncover the jets and float. Remove the drain plug.
2 Slide out the pivot pin and lift off the float. The needle valve seat is now free to be removed.
3 Remove the pilot jet, followed by the needle jet and the needle valve assembly. The main jet can be removed from the needle jet, if necessary. If required, remove the air screw and spring, noting the setting by counting the number of turns required to remove it.
4 The cold start plunger and adjuster can be removed from the cable by sliding it off sideways.
5 The throttle slide is a little unusual since it also includes a throttle stop device. Remove the throttle stop split pin, then free the rod. Remove the throttle cable by pushing the cable down the slide and moving the nipple to one side. The needle is held in the slide by a seat and spring clip. Pull the former out and remove the needle. The needle is held in position by a circlip in one of its five grooves. Note which groove holds the circlip.
6 Check that the needle is not bent by rolling it on a flat surface. If it has worn, obtain a new replacement and also renew the needle jet.
7 Check that the throttle slide is not worn or scored. Renew as necessary. If the throttle slide is badly scored, check also the condition of the carburettor body, which may also require renewal.
8 Do not use wire or any other thin metal object to clear a blocked jet, The hole can easily become enlarged or misshapen, which will seriously affect the flow of fuel. To clean the jets, blow then out with compressed air, eg; a foot pump.
9 Check the float assembly for leaks by shaking it. If petrol can be heard inside a float a new one will be required. Check the float height by measuring the distance from the gasket face to the bottom of the float assembly The float tongue must be just touching the float needle when measuring. If necessary bend the tongue to obtain the correct setting.

6 Carburettor: reassembly

1 Reassembly should be tackled in the reverse order of that given for dismantling, bearing the following points in mind.
2 Always ensure that each component is scrupulously clean before starting reassembly. Never force parts together. The diecast body of the carburettor is very prone to cracking. Do not over-tighten jets, they are of brass, and the threads are easily stripped.
3 When refitting the valve assembly, lubricate it with light machine oil.

4 Before use, check for leaks, and check that the settings are in order as described in the following Section. If dismantling was necessary because of contaminated fuel, remove and flush the fuel tank, and clean the filter in the tap. Water can cause persistent and erratic running faults, and can form due to condensation in the tank.

7 Carburettor: checking the settings

1 The various jet sizes, throttle valve cutaway and needle position are predetermined by the manufacturer and should not require modification. Check with the Specifications list at the beginning of this Chapter if there is any doubt about the sizes fitted.
2 Slow running is controlled by a combination of the throttle stop and pilot jet settings. Adjustment should be carried out as explained in the following Section. Remember that the characteristics of the two-stroke engine are such that it is extremely difficult to obtain a slow, reliable tick-over at low rpm. If desired, there is no objection to arranging the throttle stop so that the engine will shut off completely when the throttle is closed. Unlike a petroil-lubricated engine, the oil used for engine lubrication is injected directly into the engine. In consequence there is no risk of the engine 'drying up' when the machine is coasted down a long incline, if the throttle is closed.
3 As a rough guide, up to $\frac{1}{8}$ throttle is controlled by the pilot jet, $\frac{1}{8}$ to $\frac{1}{4}$ by the throttle valve cutaway, $\frac{1}{4}$ to $\frac{3}{4}$ throttle by the needle position and from $\frac{3}{4}$ to full by the size of the main jet. These are only approximate divisions, which are by no means clear cut. There is a certain amount of overlap between the various stages.
4 The normal setting of the pilot jet screw is approximately one and one half complete turns out from the closed position. If the engine 'dies' at low speed, suspect a blocked pilot jet.
5 Guard against the possibility of incorrect carburettor adjustments which will result in a weak mixture. Two-stroke engines are very susceptible to this type of fault, causing rapid overheating and often subsequent engine seizure. Changes in carburation leading to a weak mixture will occur if the air cleaner is removed or disconnected, or if the silencer is tampered with in any way. Above all, do not add oil to the petrol, in the mistaken belief that it will aid lubrication. The extra oil will only reduce the petrol content by the ratio of oil added, and therefore cause the engine to run with a permanently weakened mixture.

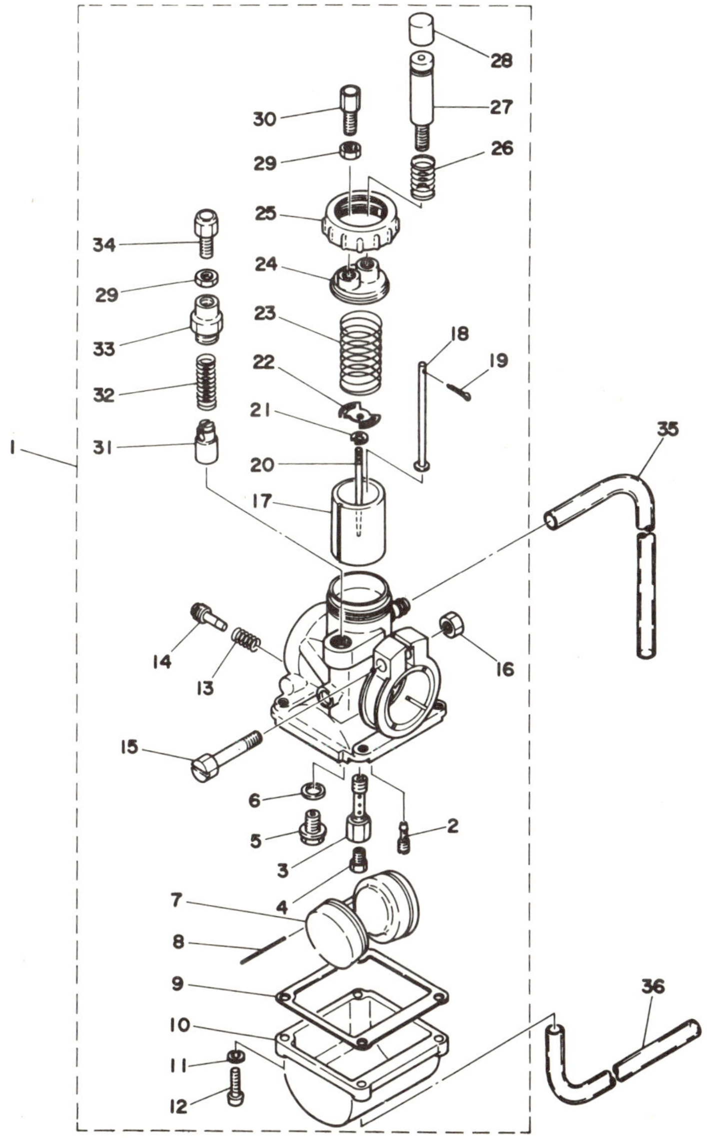

Fig. 2.2 Carburettor

1 Carburettor assembly
2 Pilot jet
3 Needle jet
4 Main jet
5 Valve seat
6 Sealing washer
7 Float
8 Pin
9 Gasket
10 Float chamber body
11 Spring washer
12 Screw – 4 off
13 Spring for air adjusting screw
14 Air adjusting screw
15 Bolt for clamp
16 Nut
17 Throttle valve (slide)
18 Throttle slide rod
19 Split pin
20 Needle
21 Circlip
22 Spring seating
23 Throttle valve return spring
24 Mixing chamber top
25 Cap
26 Throttle stop spring
27 Throttle stop screw
28 Cap for throttle stop screw
29 Cable adjuster nut – 2 off
30 Cable adjuster
31 Choke plunger
32 Plunger spring
33 Cap
34 Cable adjuster
35 Air vent pipe
36 Overflow pipe

5.1 Float bowl is secured by four screws

5.2a Displace pivot pin and remove float assembly

5.2b Needle can now be removed for examination ...

5.2c ... as can the valve seat – note sealing washer

5.3a Main jet screws into base of needle jet ...

5.3b ... which in turn screws into carburettor

5.3c Pilot jet will be found skulking in bottom of bore

5.3d Note pilot screw setting before removal

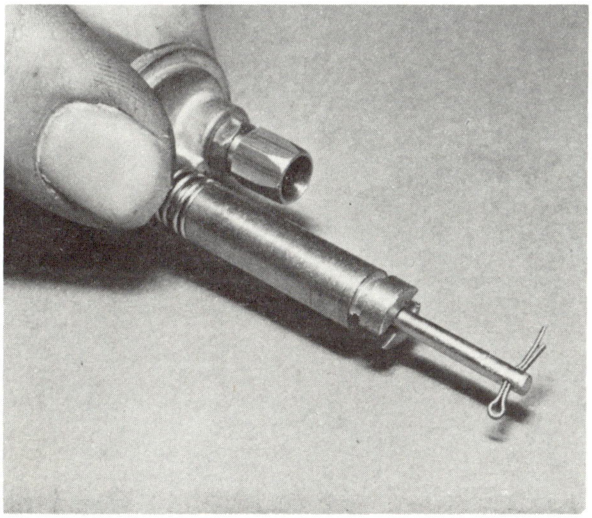

5.5a Throttle stop rod is secured by a small split pin ...

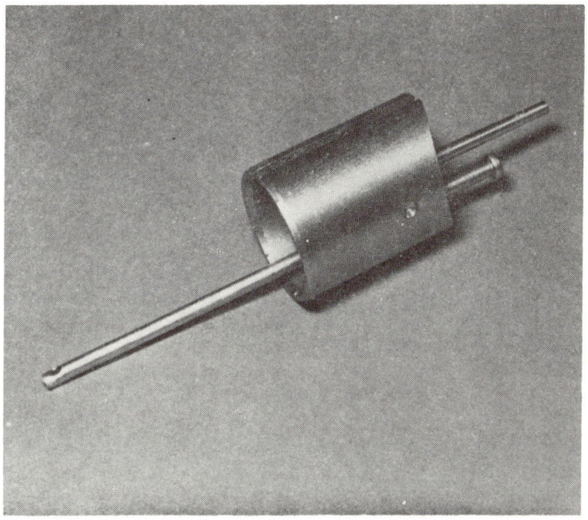

5.5b ... and may be displaced after this has been removed

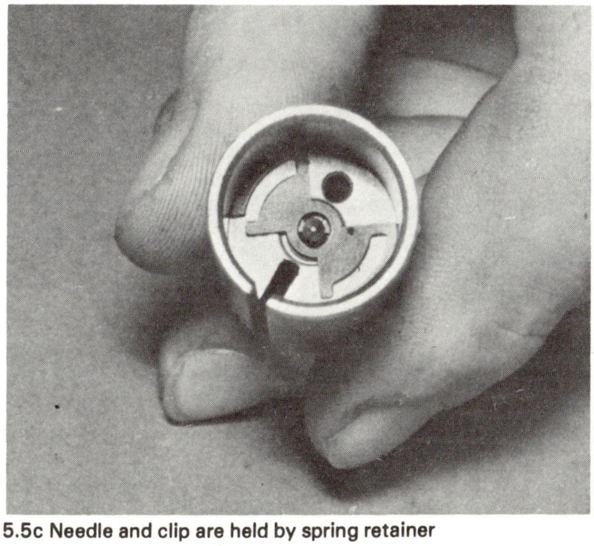

5.5c Needle and clip are held by spring retainer

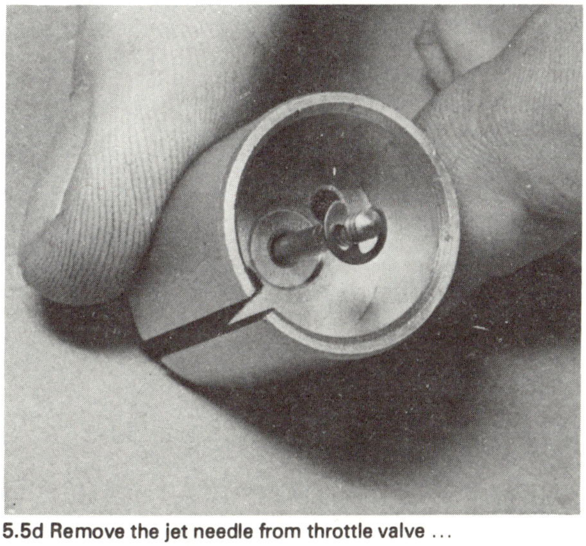

5.5d Remove the jet needle from throttle valve ...

5.5e ... noting position of clip on needle grooves

5.7 Check carburettor bore for scoring or wear

8 Carburettor: adjustment

1 Before attempting any adjustment, the engine should be run to attain normal working temperature. Two adjustments are provided; the throttle stop screw, which is situated above the carburettor, and which protrudes through the rubber cover, and the pilot mixture screw, which is located at the front of the instrument at the bottom of the choke.

2 Commence adjustment by turning the throttle stop screw to obtain a fairly fast tickover speed, and screw the pilot mixture screw fully inwards. Set the throttle stop screw to give the slowest possible reliable tickover speed. Gradually turn the pilot mixture screw anticlockwise, noting the effect which it will have on the engine speed. The screw should be set at the position which gives the fastest tickover speed. Reset the throttle stop screw to give a slow, even tickover. Note that the normal pilot screw setting is $1\frac{1}{2}$ turns out. ($1\frac{1}{4}$ on YB 100E).

9 Disc valve induction system: examination and renovation

1 As mentioned in Section 1 of this Chapter, a disc valve, or rotary valve, arrangement is employed to give more precise control of the induction timing. This gives the advantages of more efficient combustion, with improved power output and fuel economy. It is most unusual for this unit to give any trouble, as it rarely shows any marked degree of wear during the normal service life of the machine. However, should any problems arise, the assembly may be removed after dismantling the clutch and primary drive. Refer to Chapter 1, Sections 11, 12 and 15 for details.

2 Wear, and the need for renewal should be fairly obvious, such as worn splines in the disc centre and on the driving boss. The fibre valve disc may require renewal if badly worn or damaged by foreign matter becoming trapped between it and the valve cover, and likewise the cover itself. Bear in mind the remarks concerning the disc timing, as this is critical to the efficient running of the engine.

10 Air filter: removal and cleaning

1 The air filter consists of a large oil-impregnated element housed in an elaborate cylindrical canister mounted above the crankcase. The element must be removed and cleaned at 2000 mile intervals.

2 Release the domed nut or screw which secures the chromium plated cover to the left-hand end of the air filter casing. Lift the cover away, and withdraw the filter element. The element should be washed in a degreasing solvent or in petrol to remove all the accumulated dust and the old oil. Squeeze out the solvent and allow the element to dry, then re-impregnate it with clean engine oil. The filter should be soaked evenly in oil, but not to the extent that it is dripping.

3 Note that the air filter element must be renewed if it has become torn or damaged in any way. Check that the intake hoses are not split or perished, renewing where necessary.

11 Engine lubrication system: maintenance and adjustment

1 As mentioned in Section 1 of this Chapter, the engine is lubricated by an oil pump mounted to the rear of the right-hand crankcase half. It is important that an adequate level of oil is maintained in the frame-mounted oil tank, and a sight glass is provided as a means of indicating that the oil level is low. A good quality two-stroke oil should be used.

2 Ensure that all connections and unions in the oil system are kept tight and free from leaks, as leakage will cause a loss of engine lubrication which can cause rapid wear or seizure. The oil pump cable adjustment should be checked every 1000 miles/1600 km, or on any occasion where over or under-lubrication is suspected.

3 Access to the pump is gained by releasing the right-hand outer cover. Before the oil pump adjustment can be checked, it is important to ensure that the throttle cables are correctly set. Start by adjusting the lower throttle cable to give 1 mm (0.04 in) free play. The upper cable, which runs between the cable splitter box and the twistgrip assembly, is set next. Make a mark on the outer edge of the flanged part of the twistgrip rubber. The cable should be adjusted so that all free play is taken up when the mark has been turned by 3 – 6 mm (0.12 – 0.24 in).

4 Attention may now be turned to synchronising the oil pump pulley with the throttle valve. This is important, as correct adjustment will ensure that the correct quality of oil is delivered at any given throttle opening. The throttle valve has a circular mark, which can be seen by looking down the bore of the instrument. Turn the throttle twistgrip until the top of the circular mark touches the top of the carburettor bore. At this point the raised index mark on the oil pump pulley should align with the plunger pin. If this is not the case, alter the setting by way of the oil pump operating cable adjuster. Open and close the throttle a few times, then recheck the setting.

5 The oil pump stroke can be adjusted, although this is not

normally necessary. If the normal synchronisation adjustment has not cured a problem of weak or rich oil delivery, the pump stroke setting should be checked. Obviously, if the engine has seized through lack of oil, this measurement should be made to ensure that the problem does not reappear.

6 Start the engine and allow it to idle. It will be seen that the pump plunger moves slowly in and out, and this will be apparent if the pump plunger adjustment plate is watched carefully. Stop the engine when the plunger has been moved fully outwards. Measure the gap between the raised boss of the pump pulley and the adjustment plate, using feeler gauges. The feeler gauge should be a light sliding fit, and should not be forced into the gap as this will give a false reading.

7 Note the reading obtained, then repeat the operation a number of times to find the maximum gap. When the gap is at its widest possible setting, it indicates that the pump is at its minimum stroke setting. The gap reading at this point should be 0.20 – 0.25 mm (0.008 – 0.010 in). If the gap does not fall between these limits, remove the nut, spring washer and adjuster plate from the end of the pump plunger, and add or subtract shims to obtain the required clearance. Shims are available in 0.3, 0.5 and 1.0 mm (0.012, 0.020 and 0.039 in) sizes. Recheck the gap after installing the new shim(s), and if necessary, repeat the adjustment operation until the setting is correct.

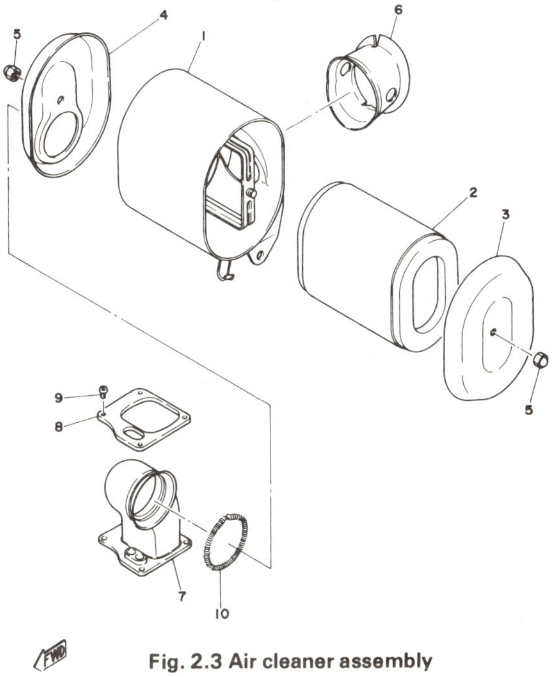

Fig. 2.3 Air cleaner assembly

1	*Air cleaner casing*	*6*	*Inlet pipe*
2	*Element*	*7*	*Adaptor*
3	*Right-hand end cover*	*8*	*Retainer plate*
4	*Left-hand end cover*	*9*	*Screw – 4 off*
5	*Nut – 2 off*	*10*	*Tension spring*

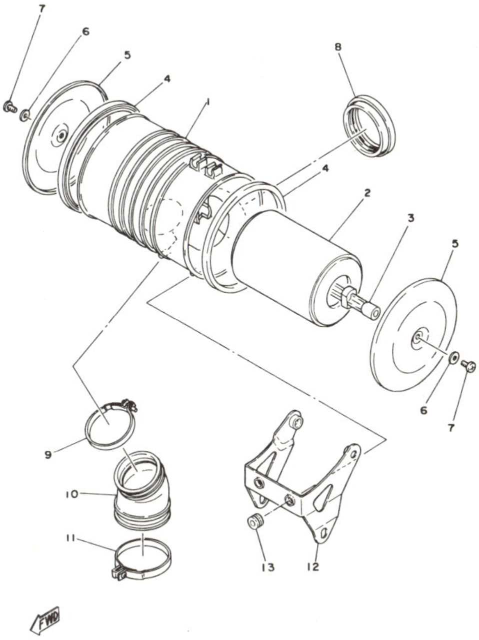

Fig. 2.4 Air cleaner assembly – later type

1	*Air cleaner case*	*8*	*Suction pipe joint*
2	*Element*	*9*	*Hose clamp*
3	*Spindle*	*10*	*Inlet pipe joint*
4	*Seal – 2 off*	*11*	*Hose clamp*
5	*Cleaner case cap – 2 off*	*12*	*Stay for air cleaner case*
6	*Gasket – 2 off*	*13*	*Grommet*
7	*Screw – 2 off*		

11.4a Circular mark on carburettor throttle valve must *just* touch top of bore ...

11.4b ... when pin and pulley mark coincide (see Fig.2.6)

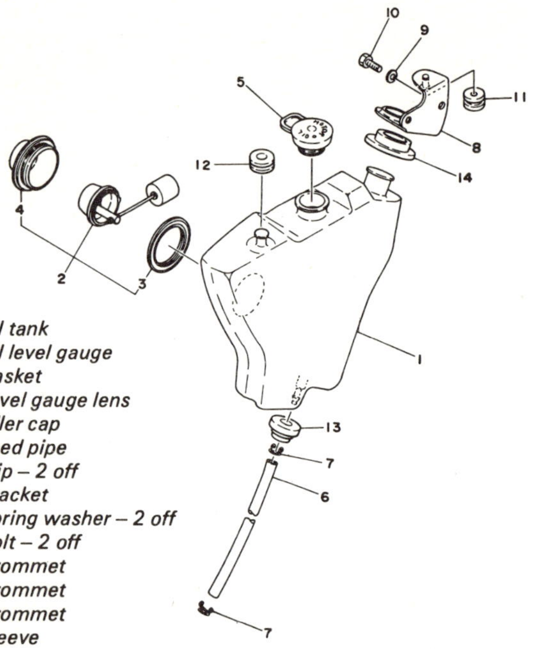

1 Oil tank
2 Oil level gauge
3 Gasket
4 Level gauge lens
5 Filler cap
6 Feed pipe
7 Clip – 2 off
8 Bracket
9 Spring washer – 2 off
10 Bolt – 2 off
11 Grommet
12 Grommet
13 Grommet
14 Sleeve

Fig. 2.5 Oil tank – later type (early models similar)

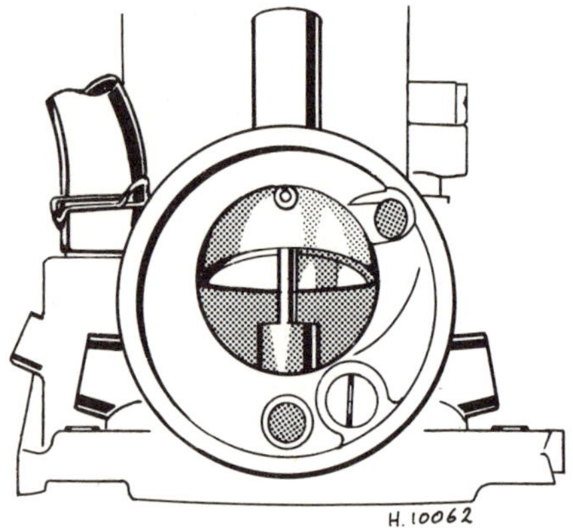

Fig. 2.6 Oil pump adjustment
Throttle valve should be positioned as shown

12 Oil pump: removal and replacement

1 The oil pump can be expected to give long service, requiring no maintenance, but in the event of failure it must be renewed. No replacement parts are obtainable, and the pump is therefore, effectively a sealed unit. The pump body is retained to the crankcase by two screws, and it can be removed after the oil feed and delivery pipes have been released.
2 When refitting the pump, use a new gasket to ensure a sound joint between the pump and the inner cover. Note that the pump cable is adjusted as described in Section 11. It will also be necessary to bleed the pump as described below.

13 Bleeding the oil pump

1 If any part of the Autolube system has been disturbed, it will be necessary to bleed the oil pump to remove any air bubbles which may have entered the pump or oil pipes. This also applies where the oil tank has been allowed to run dry. This operation is essential, as air in the pump system can cause complete loss of lubrication, a situation which is rapidly followed by engine seizure. If there is any doubt about air having entered the system, play safe and carry out the bleeding operation as a precaution.
2 Bleeding is carried out in two stages; the first being to remove any air between the oil tank and the pump body. This is a simple matter of removing the pump bleed screw, (identified by its sealing washer), and allowing the oil to flow through until any air bubbles have dissappeared. When the emerging oil is seen to be completely free of bubbles, refit the bleed screw, using a new sealing washer where necessary, and tighten the screw securely.
3 To clear the distribution part of the pump, and the delivery line, start the engine, and allow it to run at a fast idle (about 2000 rpm). Set the pump at maximum stroke by pulling the pump cable fully outwards. This will ensure the oil is delivered at its maximum rate, taking any air with it. Holding this position, let the engine run for about two minutes. The pump and lines can now be considered to be free of air.

14 Silencer: decarbonisation

1 After a considerable mileage has been covered, it will become necessary to remove the silencer baffles to clean off the accumulated oily carbon deposits. Any two-stroke engine is very susceptible to this fault, which is caused by the oily nature of the exhaust gases. As the sludge builds up back pressure will increase with resulting fall off in performance.
2 There is no necessity to remove the exhaust system in order to gain access to the baffles. They are retained in the end of the silencer by a screw, reached through a hole cut in the underside of the silencer body, close to the end. When the screw is removed, the baffles can be withdrawn.
3 If the build up of carbon and oil is not too great a wash with a petrol/paraffin mix will probably suffice as the cleaning medium. Otherwise more drastic action will be necessary such as the application of a blowlamp flame to burn away the accumulated deposits. Before the baffles are refitted they must be thoroughly clean, with none of the holes obstructed.
4 When replacing the baffles, make sure the retaining screw is located correctly and tightened fully. If the screw falls out the baffles will work loose, creating excessive exhaust noise accompanied by a marked fall off in performance.
5 Do not run the machine without baffles in the silencer or modify the baffles in any way. Although the changed exhaust note may give the illusion of greater power, the chances are that the performance will fall off, accompanied by a noticeable lack of acceleration. The carburettor is jetted to take into account the fitting of silencers of a certain design and if this balance is disturbed the carburation will suffer accordingly.

12.1a Carburettor must be released to gain access to union

12.1b Disconnect oil pipes, then remove pump

12.1c Pump worm is driven by pinion inside casing

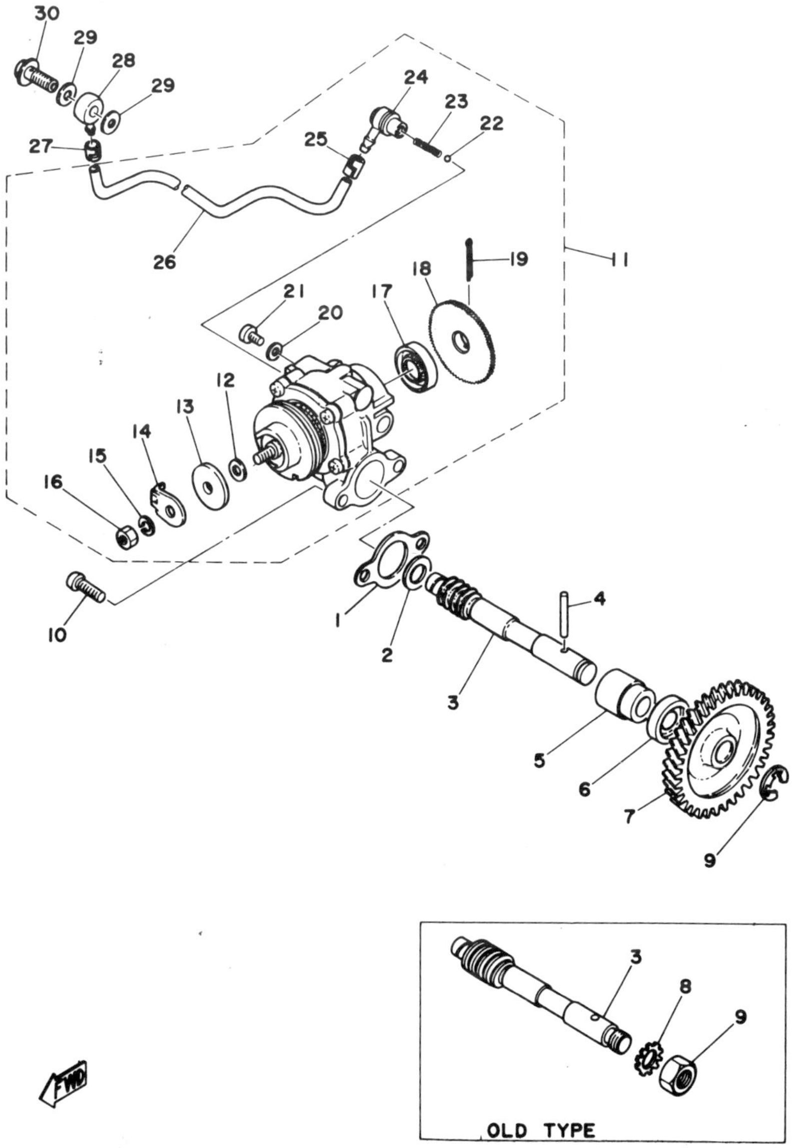

Fig. 2.7 Oil pump

1 *Pump case gasket*
2 *Washer*
3 *Worm shaft*
4 *Dowel pin*
5 *Worm shaft outer bearing*
6 *Oil seal*
7 *Drive pinion*
8 *Tooth washer*
9 *Nut*
10 *Screw – 2 off*
11 *Oil pump assembly*
12 *Shim*
13 *Adjusting plate*
14 *Adjusting plate cover*
15 *Spring washer*
16 *Nut*
17 *Oil seal*
18 *Starter plate*
19 *Split pin*
20 *Breather gasket*
21 *Bleed screw*
22 *Ball*
23 *Spring*
24 *Union*
25 *Clip*
26 *Delivery pipe*
27 *Clip*
28 *Banjo union*
29 *Banjo union gasket*
30 *Banjo union bolt*

13.2 The pump bleed screw – note sealing washer

14.1 Silencer is secured by nut and stud to frame

15 Fault diagnosis: fuel system

Symptom	Cause	Remedy
Excessive fuel consumption	Air cleaner choked or restricted Fuel leaking from carburettor Badly worn or distorted carburettor Carburettor settings incorrect	Clean or renew elements. Check all unions and gaskets. Renew. Readjust. Check settings with specifications.
Idling speed too high	Throttle stop screw in too far Carburettor top loose	Adjust screw. Tighten.
Engine sluggish. Does not respond to throttle	Back pressure in silencer	Check baffles and clean if necessary.
Engine dies after running for a short while	Blocked vent hole in filler cap Dirt or water in carburettor	Clean. Remove and clean.
General lack of performance	Weak mixture; float needle sticking in seat Air leaks at carburettor, or leaking crankcase seals.	Remove float chamber, and check needle seating. Check for air leaks or worn seals.

16 Fault diagnosis: lubrication system

Symptom	Cause	Remedy
White smoke from exhaust	Too much oil	Check oil pump setting and reduce if necessary.
Engine runs hot and gets sluggish when warm	Too little oil	Check oil pump setting and increase if necessary.
Engine runs unevenly, not particularly responsive to throttle openings	Intermittent oil supply	Bleed oil pump to displace air in feed pipes.
Engine dries up and seizes	Complete lubrication failure	Check for blockages in feed pipes, also whether oil pump drive has sheared.

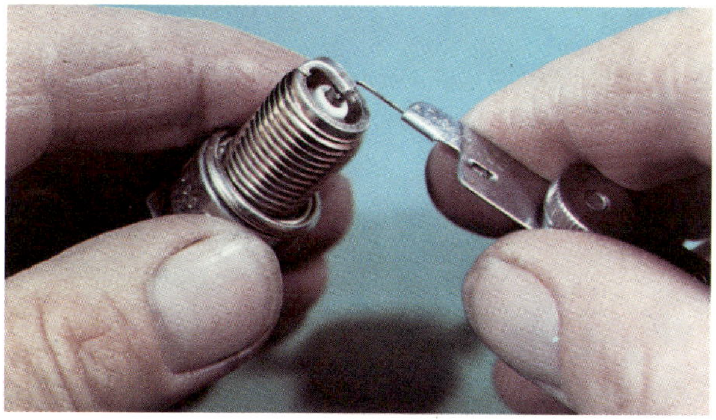

Spark plug maintenance: Checking plug gap with feeler gauges

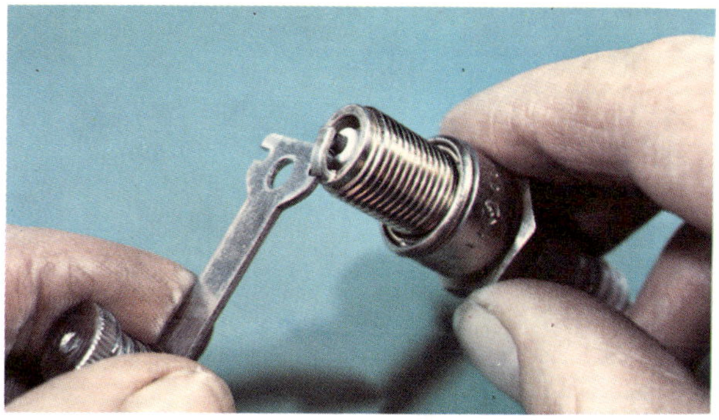

Altering the plug gap. Note use of correct tool

Spark plug conditions: A brown, tan or grey firing end is indicative of correct engine running conditions and the selection of the appropriate heat rating plug

White deposits have accumulated from excessive amounts of oil in the combustion chamber or through the use of low quality oil. Remove deposits or a hot spot may form

Black sooty deposits indicate an over-rich fuel/air mixture, or a malfunctioning ignition system. If no improvement is obtained, try one grade hotter plug

Wet, oily carbon deposits form an electrical leakage path along the insulator nose, resulting in a misfire. The cause may be a badly worn engine or a malfunctioning ignition system

A blistered white insulator or melted electrode indicates over-advanced ignition timing or a malfunctioning cooling system. If correction does not prove effective, try a colder grade plug

A worn spark plug not only wastes fuel but also overloads the whole ignition system because the increased gap requires higher voltage to initiate the spark. This condition can also affect air pollution

Chapter 3 Ignition system

Contents

General description	1	Condenser: testing	5
Contact breaker and timing adjustment	2	Ignition coil: location and testing	6
Contact breaker assembly: removal and renewal	3	Sparking plug: checking and resetting the gap	7
Condenser: location, checking and renewal	4	Fault diagnosis: ignition system	8

Specifications

Ignition system	Coil and contact breaker, direct from source coil in generator	
Ignition source coil		
Coil resistance:		
YB 100	1.72 ohms ± 10% @ 20°C	
YB 100 E	1.80 ohms ± 10% @ 20°C	
Ignition coil		
Primary winding resistance	1.6 ohms ± 10% @ 20°C	
Secondary winding resistance	6.6 k ohm ± 20% @ 20°C	
Condenser		
Capacity	0.3 microFarad	
Contact breaker		
Gap	0.3 – 0.4 mm (0.012 – 0.016 in)	
Spring pressure	0.85 ± 0.15 kg (1.874 ± 0.0331 lb)	
Ignition timing	1.8 ± 0.15 mm (0.071 ± 0.006 in) – piston position BTDC	
Sparking plug		
Make	*NGK	Motorcraft
Type	*B-7HS	AE2

*Manufacturer's recommendation

1 General description

The Yamaha YB100 models make use of a simple ac ignition system. An ignition source coil is incorporated in the flywheel generator, and is used solely to power the ignition system. A contact breaker assembly is mounted inside the flywheel generator, and is operated by a cam profile ground on the flywheel boss. Power from the source coil is fed to the contact breaker assmbly, and when the contacts are closed, they provide an easy path to earth for the generated current.

As the contacts seiarate, a sudden pulse of energy is transferred to the primary windings of the external ignition coil. The pulse in turn induces a spark in the secondary windings, and this is fed to the sparking plug.

A condenser, or capacitor, is connected in parallel with the contact breaker set, its function being to damp out any tendency towards arcing between the contact points. As is normal with small two-stroke engines, no form of advance mechanism is fitted, its use being considered unnecessary.

2 Contact breaker and timing adjustment

1 Access to the flywheel rotor and contact breaker assembly is gained after removing the circular inspection cover on the left-hand side of the engine unit. Alternatively, the left-hand outer cover can be removed to give better access. The contact breaker assembly is visible through the apertures in the flywheel rotor face. Turn the rotor until the contacts separate, and examine their faces for signs of pitting or burning. If they are obviously damaged or worn, it will be necessary to remove the flywheel rotor and to renew the contact breaker assembly. If this is the case, refer to Section 3 for further details. If the contact faces are dirty, but undamaged, they can be cleaned by drawing a strip of thin card between them.

2 No provision is made for significant changes to be made in the ignition timing setting, although small adjustments can be made by varying the contact breaker gap within its two limits. As two-stroke engines are particularly sensitive to ignition timing settings, it is important to ensure that this adjustment is

maintained accurately. To check and set the timing, it is essential to use a dial test indicator (DTI) or dial gauge. These can be obtained as a Yamaha special tool (No 90890–03002) or maybe purchased from a tool retailer, many of whom advertise in the motorcycle press. It must be decided whether the outlay on the dial gauge can be justified. If the machine is to be kept for any length of time, it is probably worthwhile. Otherwise, the checking and setting of the timing and contact breaker gap must be entrusted to a Yamaha Service Agent.

3 In addition to a dial gauge, some means of identifying the precise point at which the contacts separate will be required. Yamaha can supply a 'point checker' for this purpose (No 90890–03064), but a pocket multimeter set on the resistance scale, or a simple battery and bulb arrangement will be just as accurate and a more practical proposition for home use.

4 Remove the sparking plug and fit the dial gauge adaptor into the sparking plug hole. Position and secure the dial gauge. It is now necessary to establish top dead centre (TDC) by turning the crankshaft whilst watching the face of the dial gauge. As the piston nears the top of its stroke, the gauge needle will revolve. As TDC is passed, the needle will stop and begin to revolve in the opposite direction. Establish the exact point at which the needle stops, and set the scale of the gauge at zero. Check that the needle does not move beyond the zero position by turning the crankshaft to and fro.

5 Remove the left-hand side panel, and trace the generator output leads. On late machines, a connector is plugged into its opposite half in the battery mounting bracket. On earlier models, individual bullet terminals are hidden away inside the frame, and it may prove necessary to remove the battery mounting bracket in order to gain access to them. Where the earlier arrangement is used, disconnect each connector, taking great care not to drop the leads down into the frame, as retrieving them is very difficult. On the later models, unplug the connector block from the plastic bracket.

6 Connect one lead from the test meter, or battery bulb arrangement, to the black/white lead from the generator, and attach the remaining lead to earth. Slowly turn the flywheel rotor in an anticlockwise direction, watching the meter needle or bulb closely as the piston ascends towards top dead centre. As the contact faces just begin to separate, the meter needle will deflect, or the bulb extinguish. Establish this point exactly, if necessary turning the rotor back and forth until the precise point is found. Note that separation should always be approached from the direction of normal engine rotation to offset the effects of backlash in the system.

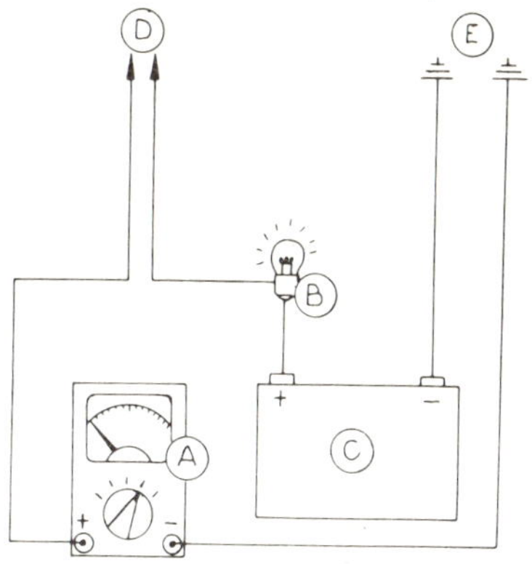

Fig. 3.1 Using multimeter or battery and bulb arrangement to establish contact breaker separation

A Multimeter set on resistance mode
B Low wattage bulb
C Battery
D Probe to moving contact terminal
E Earth

2.1 Contact breaker assembly is visible through slots in rotor

2.5 Late models have plug-in connector from generator

2.8a Check that contact breaker gap is correct after timing

2.8b A few spots of oil on wick will reduce wear

7 Check the reading on the dial gauge at the exact point of separation. The gauge should read 1·8 ± 0·15 mm (0·07 ± 0·006 in) if the timing is correct. Thus, the gauge must read between 1·95 mm and 1·65 mm (0·077 – 0·065 in). If the dial gauge reading does not fall between these limits, slacken the fixed contact securing screw, and adjust the gap setting until the correct timing setting is obtained. The timing is advanced by opening the gap and vice versa.
8 Having checked and reset the ignition timing as described above, turn the rotor until the contact breaker gap is at its widest, and check this using feeler gauges. The nominal setting is 0·35 mm (0·014 in), but this can be varied between 0·30 mm and 0·40 mm (0·012 – 0·016 in) if the ignition timing setting demands this. If the gap is outside these limits after the timing has been set, it will be necessary to purchase and fit a new contact breaker set to achieve both timing and contact breaker gap settings. The normal reason for this is wear of the fibre heel of the contact set. It follows that it is worthwhile placing one or two drops of oil on the felt lubricating wick to minimise the effects of wear. When the new contact breaker set is fitted, it will be necessary to set the gap and then recheck the accuracy of the ignition timing.

3 Contact breaker assembly: removal and renewal

1 If the contact breaker points are found to be burnt or pitted, or if the assembly has worn so that it is no longer possible to achieve an acceptable timing and gap setting, it will be necessary to remove the assembly and to fit a new replacement set. Access to the contact breaker assembly is gained after removing the left-hand outer cover, and withdrawing the flywheel rotor. This latter operation will necessitate the use of a Yamaha extractor, No 90890–01189. The procedure for removing the flywheel rotor is described in Chapter 1, Section 16, paragraphs 1 to 3.
2 The contact breaker assembly is secured to the generator stator by a single screw. Before the assembly can be removed, it will be necessary to detach the coil and condenser lead from the spring blade of the moving point. They are retained by a nut, which should be replaced to avoid displacing or losing the two insulating washers, after the spring blade has been removed from the support post.
3 Using a small electrical screwdriver, prise off the circlip which retains the moving contact assembly to its pivot pin. Remove the plain washer, followed by the moving contact complete with insulating washers. Make a note of the order in which components are removed, as they are easily assembled incorrectly.
4 Reassemble the contact breaker assembly by reversing the dismantling sequence, taking care that the insulating washers are replaced correctly. If this precaution is not observed, it is easy to inadvertently earth the assembly rendering it inoperative. The pivot pin should be greased sparingly, and a few drops of oil applied to the cam lubricating wick.
5 If the contact breaker is being renewed due to excessive burning of the contacts, this is likely to have been caused by a faulty condenser. Refer to the next Section if this is suspected.

4 Condenser: location, checking and renewal

1 A condenser is included in the contact breaker circuit to prevent arcing across the contact faces as they separate. It is connected in parallel with the contact set, and if the condenser fails in service, ignition problems will undoubtedly arise.
2 If the engine is difficult to start, or if misfiring occurs, it is possible that the condenser is at fault. To check whether the condenser has failed, observe the points whilst the engine is running, after removing the circular portion of the left-hand

3.2 Contact breaker assembly is secured by screw

4.3 Condenser is mounted on generator stator

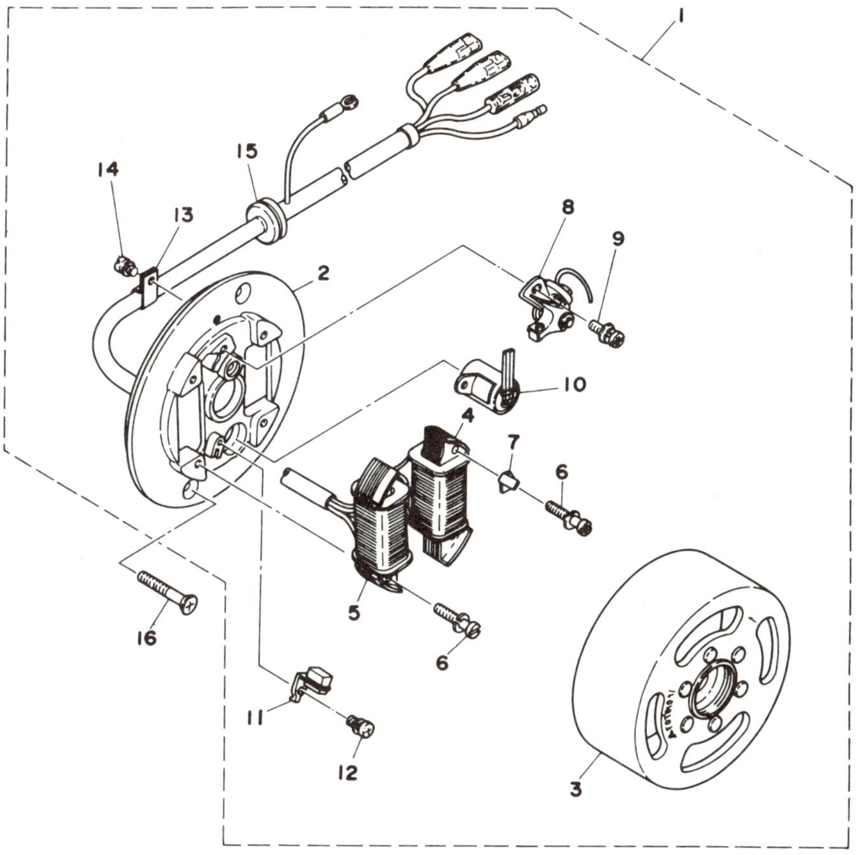

Fig. 3.2 Flywheel generator

1 *Flywheel generator assembly*
2 *Stator*
3 *Rotor*
4 *Lighting/charge coil*
5 *Ignition source coil*
6 *Screw – 4 off*
7 *Pointer*
8 *Contact breaker assembly*
9 *Screw*
10 *Condenser*
11 *Lubricating wick*
12 *Screw*
13 *Cable clamp*
14 *Screw*
15 *Grommet*
16 *Screw – 2 off*

crankcase cover. If excessive sparking occurs across the contact points and they have a blackened or burnt appearance, it may be assumed the condenser is no longer serviceable.

3 The condenser is attached to the inside of the stator cover, and is retained by a single screw through the strap soldered to the body of the condenser and by the lead wire attached to the screw and nut passing through the end of the moving contact return spring. Remove the screw and nut so that the terminal end is freed. Because it is impracticable to repair a defective condenser, a new one must be fitted.

5 Condenser: testing

1 Without the appropriate test equipment, there is no alternative means of verifying whether a condenser is still serviceable.

2 Bearing in mind the low cost of a condenser, it is far more satisfactory to check whether it is malfunctioning by direct replacement.

6 Ignition coil: location and testing

1 Ignition coils normally have a long and trouble-free working life. If, however, a persistent ignition fault cannot be resolved after checking the contact breaker condition and setting, the condenser and the ignition timing, it is possible that the coil has broken down. In many cases, the coil will not fail completely, but will operate inconsistently.

2 A limited number of tests may be performed at home. Start by removing the sparking plug cap, and arranging the bared high tension lead about 5 mm from the cylinder head fins. Switch on the ignition, and spin the engine over using the kickstart. It should be noted that the high tension spark will take the quickest route to earth, whether it is by jumping across to the cylinder head fin or by way of a nearby owner's finger. Although any shock received is not dangerous, it is rather startling, and is best avoided by lodging the lead in position whilst this test is made.

3 A fat bluish-white spark should be seen when the engine is spun, indicating that the system is functioning well. If the spark is thin and yellowish, it is likely that the ignition coil is failing. Check that the suppressor cap and sparking plug are in sound condition by repeating the check with them in position. Elusive faults can often be traced to these two components.

4 For those owners possessing a pocket multimeter, a number of further tests can be made. Trace the high tension lead back to the ignition coil, and disconnect the black/white lead at the coil connection. With the meter set on the ohms scale, connect the positive (+) lead to the black white lead from the coil, and the negative lead to a sound earth point. A reading of 1·6 ohms ± 10% (1·44 – 1·76 ohms) should be obtained, and is known as the primary winding resistance. Connect the positive probe lead to the end of the HT lead, the negative lead remaining earthed. Set the meter on kilo ohms. The meter should indicate a secondary winding resistance of 6·6 K ohms ± 20% (5·28 – 7·92 K ohms).

5 If the above tests give readings significantly outside those shown above, the coil can be considered defective. Borderline readings should always be cross-checked by having the coil tested professionally before the unit is condemned. Open or short circuits will be obvious when checked with the meter, and are a definite indication that the coil is defective. Further testing can only be carried out on expensive and specialised test equipment. For this reason, this type of work must be entrusted to a Yamaha Service Agent or an auto-electrical specialist.

6 Another possible source of problems is the ignition source coil, which is located in the flywheel generator. This can be tested in position by measuring its resistance. Connect the positive (+) probe lead to the black/white generator output lead, and the negative (–) probe lead to earth. Set the meter on the ohms scale. The following reading should be obtained.

1·8 ohms ± 10% at 20°C

If outside these limits, the source coil should be considered unserviceable, and must be renewed.

7 Sparking plug: checking and resetting the gap

1 Yamaha fit an NGK B–7HS sparking plug to the models, as standard equipment. The recommended sparking plug gap is 0·6 – 0·7 mm (0·024 – 0·028 in).

2 The gap should be checked and re-set every 1000 miles as part of the normal service. In addition, in the event of a roadside breakdown where the engine has mysteriously 'died', the sparking plug should be the first item checked. To re-set, bend the outer electrode away from or closer to the centre electrode and check that a 0·6 mm feeler gauge can be inserted. Never bend the central electrode, otherwise the insulator will crack, causing engine damage if the broken particles fall in whilst the engine is running.

3 After some experience the sparking plug electrodes can be used as a reliable guide to engine operating conditions. See the colour photographs on plug condition. Refer to page 67.

4 Always carry a spare sparking plug of the correct type. The plug in a two-stroke engine leads a particularly hard life and is liable to fail more readily than when fitted to a four-stroke.

5 Never overtighten a sparking plug, otherwise there is risk of stripping the threads from the cylinder head, especially as it is cast in light alloy. A stripped thread can be repaired without having to scrap the cylinder head by using a 'Helicoil' thread insert. This is a low-cost service, operated by a number of dealers.

6 Before replacing a sparking plug into the cylinder head coat the threads sparingly with a graphited grease to aid future removal. Use the correct sized spanner when tightening a plug, otherwise the spanner may slip and damage the ceramic insulator. The plug should be tightened sufficiently to seat firmly on its sealing washer, and no more.

7 Make sure that the plug insulating cap is a good fit and free from cracks. Apart from acting as an insulator from water and road dirt it contains the suppressor for eliminating radio and TV interference.

8 Fault diagnosis: ignition system

Symptom	Cause	Remedy
Engine will not start	No spark at plug	Faulty ignition switch. Check whether current is reaching ignition coil.
	Weak spark at plug	Dirty contact breaker points require cleaning. Contact breaker gap has closed up. Re-set.
Engine starts, but runs erratically	Break or short in LT circuit	Locate and rectify. If no improvement check whether points are arcing. If so renew condenser.
	Ignition timing incorrect	Check ignition timing and if necessary, reset. Renew contact breaker set if required.
	Plug lead insulation breaking down	Check for breaks in outer covering, especially near frame.
Engine difficult to start and runs sluggishly. Overheats	Ignition timing retarded	Check ignition timing and advance to correct setting. Renew contact breaker set if required.

Chapter 4 Frame and forks

Contents

General description	1
Front forks: removal – general	2
Front fork legs: removal from yokes	3
Fork yokes: removal and replacement	4
Fork legs: dismantling	5
Front forks: examination and renovation	6
Front forks: reassembly and replacement	7
Steering head lock: location and examination	8
Frame: examination and renovation	9
Swinging arm rear fork: removal, examination and renovation	10
Rear suspension units: examination	11
Centre stand: examination	12
Prop stand: examination	13
Footrests: examination	14
Rear brake pedal: examination and renovation	15
Kickstart lever: examination and renovation	16
Speedometer head: location and examination	17
Speedometer drive cable: examination and maintenance	18
Speedometer drive: location and examination	19
Cleaning the machine	20
Fault diagnosis: frame and forks	21

Specifications

Frame

Type	Pressed-steel spine

Steering head

Castor angle	26° 30′
Trail	81 mm (3.216 in)
Steering head bearings:	
Upper	19 x $\frac{1}{4}$ in steel balls
Lower	19 x $\frac{1}{4}$ in steel balls

Front forks

Type	Oil damped telescopic
Travel	100 mm (3.94 in)
Spring free length	165.5 mm (6.52 in)
Stanchion diameter	27 mm (1.06 in)
Fork oil capacity	165 $\pm$ 4 cc (10.07 $\pm$ 0.24 cu in)
Fork oil grade	SAE 15 motor oil or fork oil

Rear suspension

Type	Oil damped swinging arm
Travel	65 mm (2.56 in)
Spring free length	198 mm (7.8 in)
Swinging arm free play	1 mm (0.04 in) maximum
Pivot shaft diameter	10 mm (0.39 in)

1 General description

The Yamaha YB 100 models feature a spine frame constructed from steel pressings welded together to form a monocoque structure to which the engine unit, forks and swinging arm assembly are appended. This method of construction combines lightness in weight with great rigidity. The spine frame has for many years been used with success on lightweight motorcycles, but the recent appearance of medium and heavyweight spine-framed roadsters is indicative of its increasing popularity.

The front forks are of the conventional oil-damped telescopic type, whilst the rear wheel is carried in a swinging arm fork, the movement of which is controlled by a pair of sealed rear suspension units.

2 Front forks: removal – general

1 It is unlikely that the forks will require removal from the frame unless the fork seals are leaking or accident damage has

been sustained. In the event that the latter has occurred, it should be noted that the frame may also have become bent, and whilst this may not be obvious when checked visually, it could prove to be potentially dangerous.

2 If attention to the fork legs only is required, it is unnecessary to detach the complete assembly, the legs being easily removed individually.

3 If, on the other hand, the headstock bearings are in need of attention, the forks complete with bottom yoke must be removed.

4 Before any dismantling work can be undertaken, the machine should be placed on the centre stand, and blocked securely so that the front wheel is held off the ground. Detach the speedometer drive cable at the wheel, by unscrewing the knurled gland nut which retains it. Remove the front brake cable at the actuating lever.

5 Remove the wheel spindle nut and split pin, and withdraw the spindle with the aid of a tommy bar. The wheel can now be lowered clear of the forks and put to one side. To avoid damage to the paintwork, it is a good idea to remove the front mudguard at this stage, irrespective of whether removal will later be necessary, as in the case of individual fork leg removal.

3 Front fork legs: removal from yokes

1 If attention to the fork legs only is required, they are easily removed without disturbing the fork yokes or steering head bearings. It may also be considered worthwhile removing the fork legs as a preliminary to dismantling the steering head assembly, as the latter will be less unwieldly with the forks removed.

2 Commence preliminary dismantling as described in Section 2 of this Chapter. Each fork leg is retained to the upper yoke by a chromiun plated bolt. These should be slackened, but not removed completely, at this stage. Slacken the clamp bolts in the lower yoke. Each leg may be removed by drawing it downwards and clear of the fork yokes. To this end, it is useful to tap the fork top bolt, using a hide mallet, to jar each leg free. Once started, remove the top bolts and pull the leg clear.

3 If the stanchion is still held firmly in the yokes, it is permissible to spread the lower yoke bore slightly, using a large screwdriver. As the fork leg assembly is withdrawn, the upper shrouds will remain in place, and need not be disturbed unless specific attention is required.

2.5 Mudguard is retained by a total of four bolts

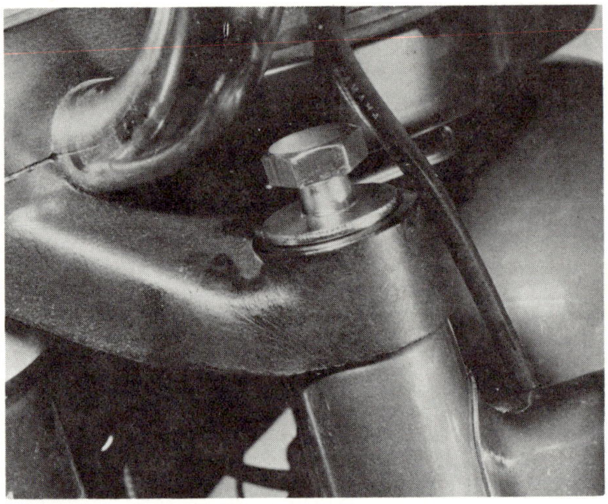

3.2a Slacken the fork top bolts ...

3.2b ... ana release clamp bolts at lower yoke

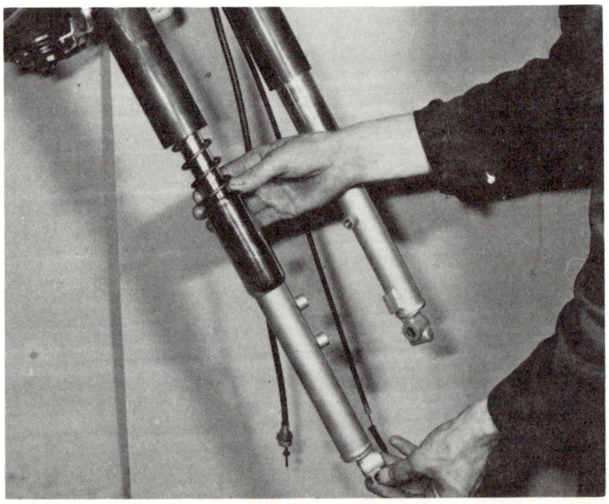

3.2c Fork leg can be pulled downwards to clear yokes

4 Fork yokes: removal and replacement

1 As mentioned previously, it is possible to remove the lower yoke with the fork legs in position, if necessary. This course of action is not recommended, as the dismantling work is made much more difficult. It is far easier if the fork legs are removed first, as described in Section 3. Before further dismantling work is undertaken, place a blanket or similar over the petrol tank to avoid damage to the paintwork.
2 Slacken the four bolts which secure the handlebar assembly by way of a clamp block (early models) or two small clamp halves (late models). The handlebar assembly can be rested on the petrol tank, unless removal is specifically required, in which case the various control cables and electrical leads must be traced and disconnected.
3 Disengage the two lower shroud sections from the lower yoke, and place them to one side. The upper shroud sections incorporate the headlamp and indicator mountings, and these components can be removed as a unit, leaving the various electrical cables attached. Disengage the assembly from the yokes, and tie it to the frame to avoid straining the electrical connections. Release the instrument pod from the underside of the top yoke, and arrange this in a similar manner.
4 Slacken and remove the chromium plated bolt which secures the top yoke to the steering stem. The top yoke can now be released, by tapping it upwards with a soft-faced mallet, and placed to one side to await reassembly
5 Using a 'C' spanner slacken and remove the slotted locknut at the top of the stem. Before removing the second of the steering stem nuts, it should be noted that the two head races contain a total of 38 steel balls, each of which will prove remarkably elusive if allowed to drop free onto a hard surface. Some provision must be made to catch them, and a large piece of cloth spread below the steering head should contain any of the balls which manage to avoid capture.
6 Lower the yoke carefully, and remove any of the steering head balls which may have stuck to the yoke or the head races. Check that all of the balls have been retrieved and place them in a closed container to avoid possible loss.
7 When reassembling the yokes and steering head, check that the bearing tracks and balls are clean and re-greased. The balls can be held in position with grease during assembly, noting that 19 are fitted in each race. Before securing the bolt which retains the upper yoke, check that the steering head bearings are correctly adjusted, and that the yokes are aligned by temporarily refitting the fork legs. Steering head bearing adjustment should take place after the handlebars and forks have been refitted, so it is advisable to leave the steering stem bolt finger tight until reassembly and adjustment have been completed.

4.2 On early models, ignition switch is incorporated in clamp

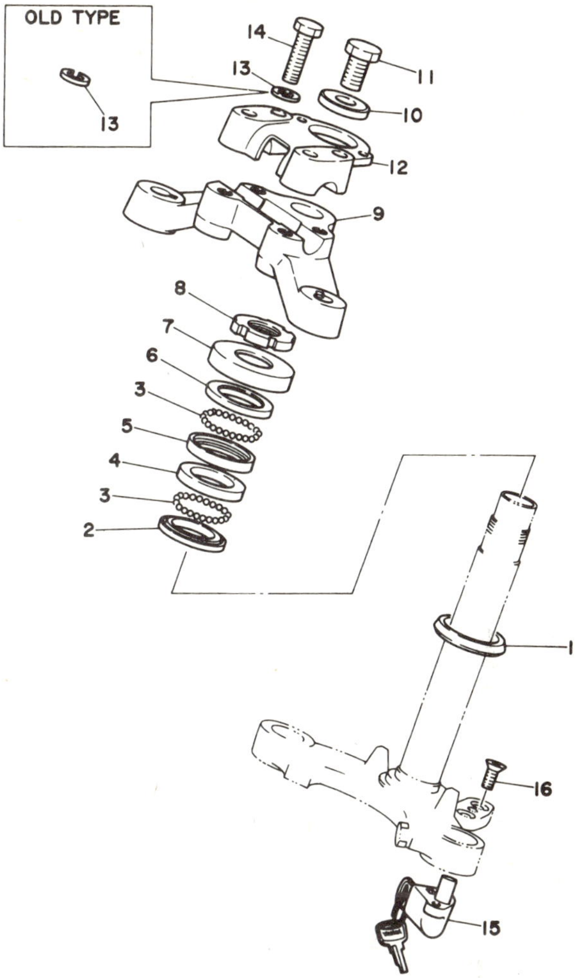

Fig. 4.1 Steering head assembly

1	*Dust seal*	9	*Top fork yoke*
2	*Lower cone*	10	*Washer*
3	*Ball bearing ($\frac{1}{4}$ inch) – 38 off*	11	*Bolt*
		12	*Upper handle bar clamp*
4	*Lower cup*	13	*Spring washer – 4 off*
5	*Upper cup*	14	*Bolt – 4 off*
6	*Upper cone*	15	*Steering column lock assembly*
7	*Ball race cover*		
8	*Adjusting nut*	16	*Screw – 2 off*

5 Fork legs: dismantling

1 With the fork leg(s) removed, the spring shroud or gaiter and external fork spring can be slid off the stanchion together with the spring seat. The stanchion is retained in the lower leg by a sleeve nut which also supports the oil seal.
2 The sleeve nut can be removed with the aid of a strap wrench, if this is available. A good alternative method is to wrap a piece of rag around the nut to protect the polished surface, and then tighten one or two worm drive hose clips around it. A large adjustable wrench can then be used to slacken the nut, bearing against the raised boss of the clips.
3 With the nut slackened, the stanchion can be withdrawn complete with the large, headed fork bush. The various components can then be slid upwards off the stanchion.

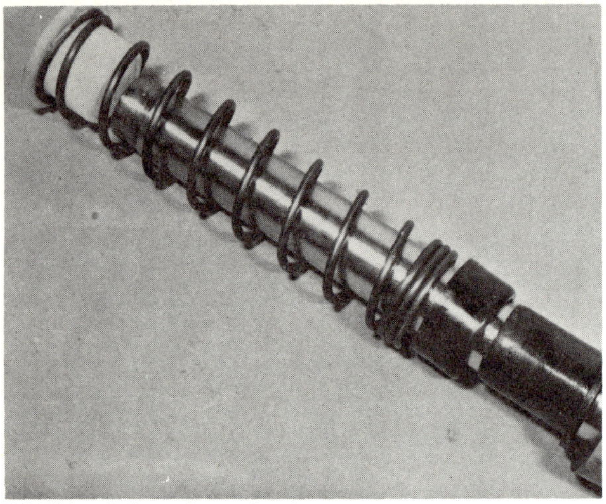

5.1 Slide off the spring and spring seats

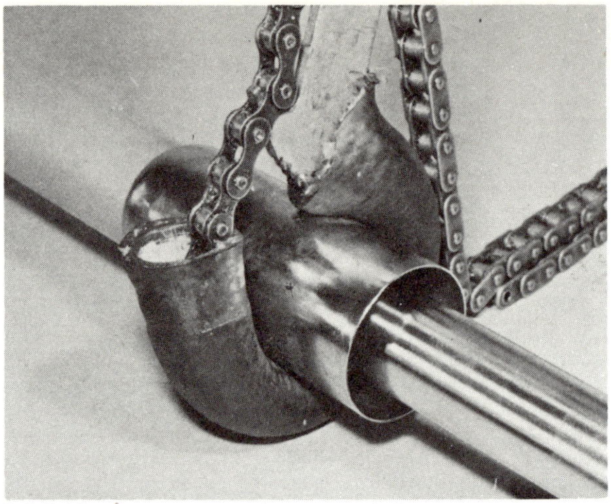

5.2 Rubber hose prevents damage to plating

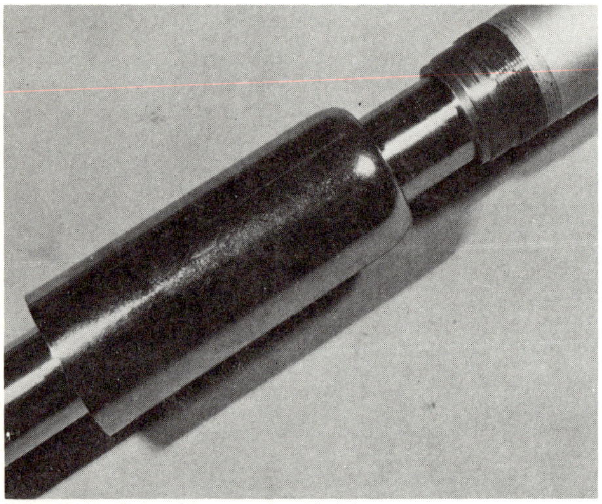

5.3a Slide off the sleeve nut/seal holder ...

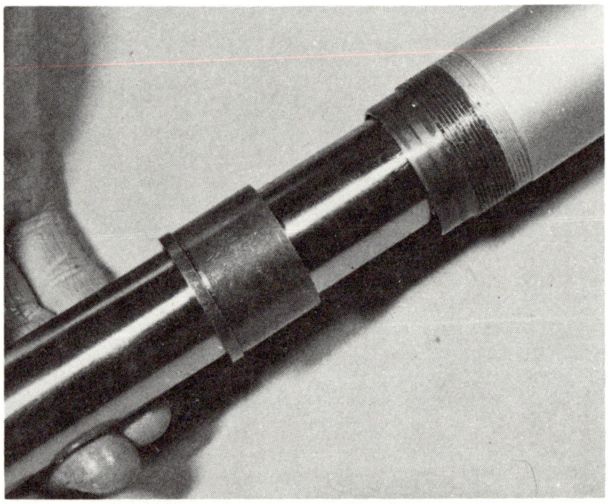

5.3b ... and withdraw stanchion and bushes

6 Front forks: examination and renovation

1 The parts most liable to wear over an extended period of service are the bushes and the oil seals, especially where gaiters are not fitted. Wear is normally accompanied by a tendency for the forks to judder when the front brake is applied and it should be possible to detect the increased amount of play by pulling and pushing on the handlebars when the front brake is applied fully. This type of wear should not be confused with slack steering head bearings, which can give identical results.
2 Renewal of the worn parts is quite straightforward. Particular care is necessary when renewing the oil seals. Both the seal and the fork tube should be greased, to lessen the risk of damage.
3 After an extended period of service the fork springs may take a permanent set. If the overall length has decreased it is wise to fit new components. Always fit new springs as a matched pair, never separately.
4 Check the outer surface of the fork stanchions for scratches or roughness. It is only too easy to damage the oil seals during reassembly, if these high spots are not eased down. The fork stanchions are unlikely to bend unless the machine is damaged in an accident. Any significant bend will be detected by eye, but if there is any doubt about straightness, roll the stanchions on a flat surface. If the stanchions are bent they must be renewed. Unless specialised repair equipment is available, it is rarely practicable to straighten them to the necessary standard.
5 The dust seals must be in good order if they are to fulfill their proper functions. Replace any that are split or damaged.
6 Damping is effected by the damper units contained within each fork stanchion. The damping action can be controlled within certain limits by changing the viscosity of the oil used as the damping medium, although a change is unlikely to prove necessary except in extremes of climate.
7 Roughness in the steering head can usually be attributed to worn or damaged steering head bearings. Examine each of the steel balls, rejecting any which show signs of pitting of corrosion. If the balls are worn, it is possible that the steering head cups and cones will also require renewal. The upper and lower cups are pressed into the steering head, and may be removed by drifting them out from the opposite side. The lower cone may be prised off the steering stem using suitable levers. When fitting the new cups, ensure that they enter the bore squarely and seat securely. When reassembling the steering head and bearings ensure that the latter are well greased to avoid subsequent problems with wear or corrosion.

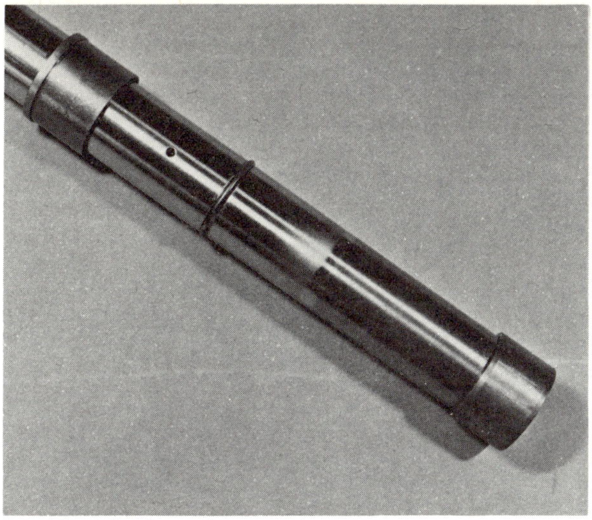

6.1a Check for signs of wear or damage

6.1b Seal should be renewed as a precaution

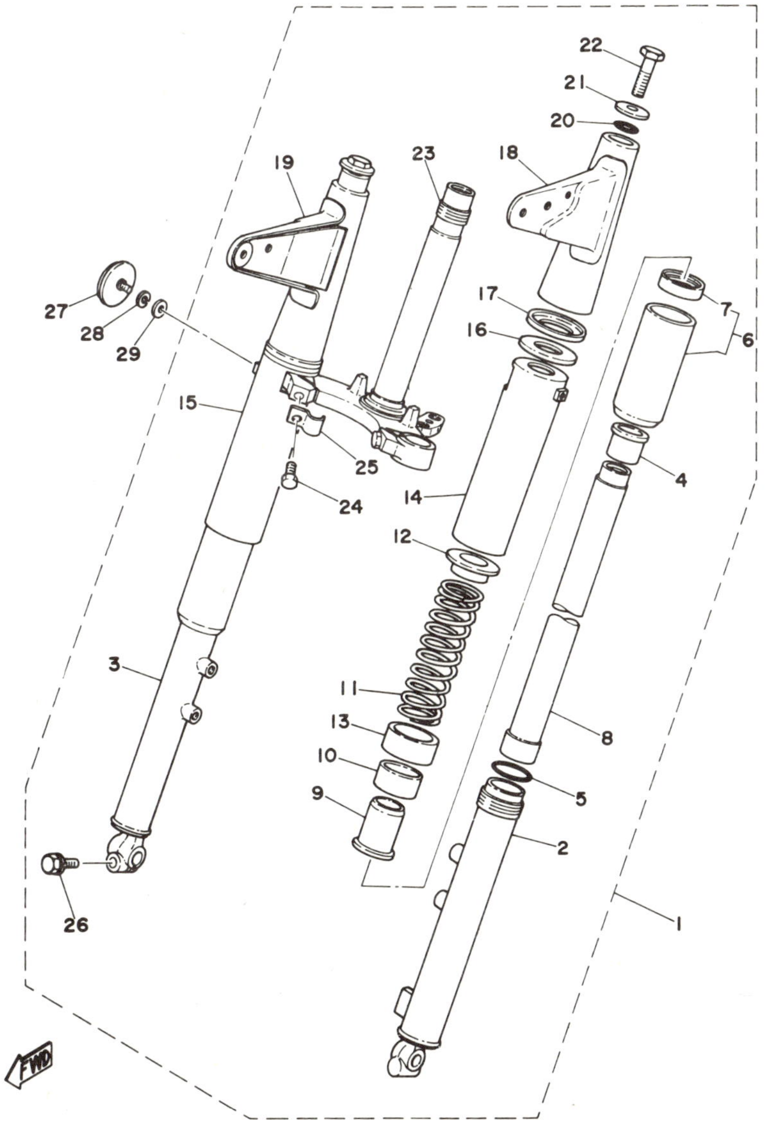

Fig. 4.2 Front forks

1 Front fork assembly
2 Left-hand lower leg
3 Right-hand lower leg
4 Fork top bush – 2 off
5 O-ring – 2 off
6 Shroud assembly – 2 off
7 Oil seal – 2 off
8 Stanchion – 2 off
9 Fork spring seating – 2 off
10 Spring guide – 2 off
11 Fork spring – 2 off
12 Spring seat – 2 off
13 Spacer – 2 off
14 Left-hand outer shroud
15 Right-hand outer shroud
16 Packing piece – 2 off
17 Top cover guide – 2 off
18 Left-hand headlamp bracket
19 Right-hand headlamp bracket
20 O-ring – 2 off
21 Washer – 2 off
22 Top bolt – 2 off
23 Lower yoke assembly
24 Bolt – 2 off
25 Cable clip
26 Bolt
27 Reflector – 2 off
28 Spring washer – 2 off
29 Washer – 2 off

7 Front forks: reassembly and replacement

1 Replace the front forks by reversing the dismantling sequence described in Sections 2, 3 and 4 of this Chapter.
2 With the components in position, but not tightened down, bounce the forks up and down a few times. This ensures that each component part of the assembled unit settles into its natural position, and suffers no subsequent strain when the assembly is tightened down.
3 Tighten the unit from the wheel spindle upwards, do not omit the split pin from the wheel spindle nut. Add the recommended quantity of fork oil. As an alternative, an SAE 15 engine oil may be used. Fork oil is to be preferred, however, as it has additives in it to suppress any tendency towards frothing.
4 On machines with fork shrouds in particular, some difficulty may be encountered in drawing the stanchion into position. The manufacturers recommend the use of an assembly tool consisting of a threaded rod with a 'T' handle which screws into the internal thread on the stanchion, enabling it to be drawn into position. A more convenient expedient is to use a suitable bolt to pull the stanchion into position. The swinging arm pivot bolt will be found to be ideal.
5 Check the adjustment of the steering head bearings before the machine is used on the road and again shortly afterwards, when they settle down. If the bearings are too slack, fork judder will occur. There should be no play at the headraces when the handlebars are pulled and pushed hard, with the front brake applied hard.
6 Overtight head races are equally undesirable. It is possible to place a pressure of several tons on the head bearings by over-tightening, even though the handlebars may seem to turn quite freely. Overtight bearings will cause the machine to roll at low speeds and give imprecise steering. Adjustment is correct if there is no play in the bearings and the handlebars swing to full lock either side when the machine is on the centre stand with the front wheel clear of the ground. Only a light tap on each end should cause the handlebars to swing.

8 Steering head lock: location and examination

1 The steering head lock is attached to the underside of the lower yoke. It is retained by screws. When in a locked position, the plunger extends and engages with a portion of the steering head stem, so that the handlebars are locked in position and cannot be turned.
2 If the lock malfunctions, it must be renewed. A repair is impracticable. When the lock is changed it follows that the key must be changed too, to correspond with the new lock.

7.4 Pivot bolt helps when refitting fork legs

7.5a Steering head bearings should be carefully set up

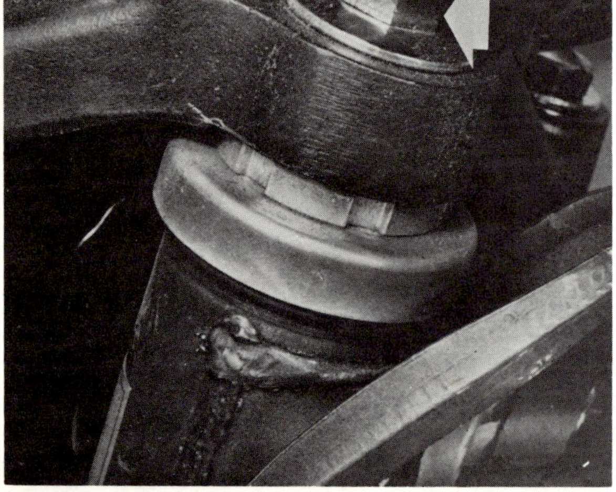

7.5b Tighten top bolt after adjustment

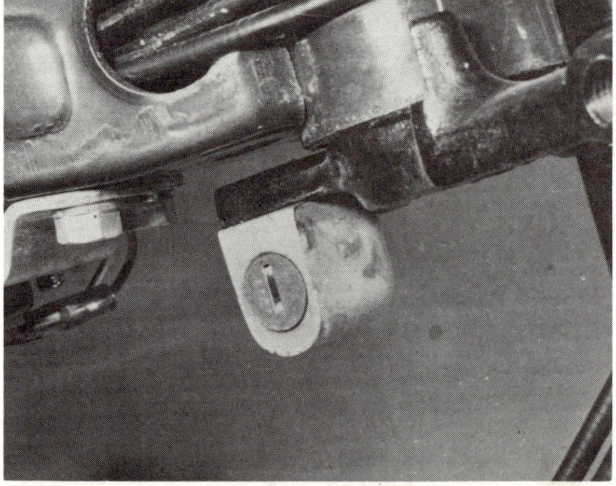

8.1 Steering lock is mounted on underside of yoke

9 Frame: examination and renovation

1 The frame is unlikely to require attention unless it is damaged as the result of an accident. In many cases, replacement of the frame is the only satisfactory course of action, if it is badly out of alignment. Comparatively few frame repair specialists have the necessary mandrels and jigs essential for the accurate re-setting of the frame and, even then there is no means of assessing to what extent the frame may have been overstressed such that a later fatigue failure may occur.

2 After a machine has covered an extensive mileage, it is advisable to keep watch for signs of cracking or splitting at any of the welded joints. Rust can cause weakness at these joints particularly if they are unpainted. Minor repairs can be effected by welding or brazing, depending on the extent of the damage found.

3 A frame out of alignment, will cause handling problems and may even promote 'speed wobbles' in a particular speed range. If misalignment is suspected as the result of an accident, it will be necessary to strip the machine so that the frame can be checked, and if need be, renewed.

10 Swinging arm rear fork: removal, examination and renovation

1 The swinging arm fork pivots on two bonded rubber bushes, which are a press fit in the fork ends. The bushes bear upon a pivot shaft which passes through the frame crossmember.

2 Place a machine on the centre stand, and secure it so that the rear wheel is raised clear of the ground. The chainguard must be released from the swinging arm. This takes the form of a complete enclosure, comprising two pressed-steel sections. These are retained by short cross-headed screws.

3 Remove the final drive chain and place it to one side. It may be worthwhile relubricating the chain at this stage.

4 Detach the rear brake torque arm at the brake plate. It is held by a spring pin, nut and washer. Unscrew the rear brake adjusting nut and free the operating lever from the rod. Place the spring, trunnion and nut back on the rod, to avoid loss.

5 Remove the split pin and castellated nut from the left-hand end of the rear wheel spindle. Do not, at this stage, disturb the large, plain nut. Using a tommy bar, or suitable substitute, withdraw the wheel spindle from the right-hand side. The wheel will

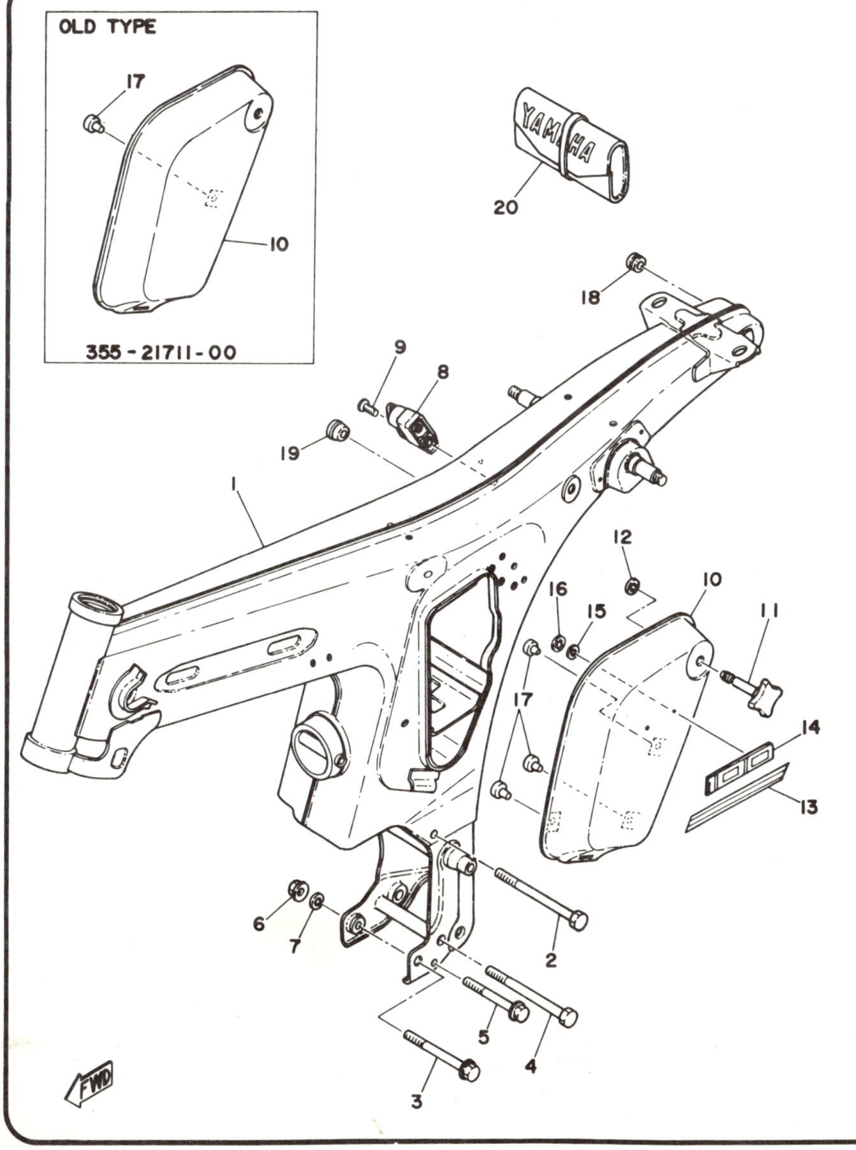

Fig. 4.3 Frame assembly

1 Frame
2 Bolt
3 Bolt
4 Bolt
5 Bolt
6 Nut
7 Spring washer
8 Helmet holder
9 Screw
10 Left-hand side cover
11 Knob for left-hand cover
12 Spring washer
13 Side cover emblem
14 Side cover emblem
15 Washer – 2 off
16 Nut – 2 off
17 Rubber for side cover
18 Grommet
19 Grommet
20 Tool roll

remain in position as it is retained by the outer, left-hand spindle. Pull out the distance piece between the brake plate and swinging arm, and also the right-hand chain adjuster. The brake plate assembly can now be removed, if desired. with the wheel and final drive assembly undisturbed.

6 Unscrew the large diameter nut on the outer spindle situated on the left-hand side. Disengage and remove the wheel complete with final drive sprocket and cush drive unit. The left-hand chain adjuster can now be lifted away.

7 Moving to the right-hand side of the machine, disconnect the spring from the brake light switch arm. Release the single screw which retains the switch to the swinging arm, and lift the switch clear. Release the electrical leads from the switch terminals, and pull the leads out of the swinging arm brace tube.

8 Unscrew the nut on the pivot shaft, leaving the shaft in position for the moment. Remove the two lower suspension unit nuts and bolt and raise the units clear of their mounting brackets. Withdraw the pivot shaft. It may be necessary to drive the shaft out using a long rod or bolt as a drift. Lift the swinging arm fork clear of the frame.

9 Examine the condition of the bushes in the fork ends. If damaged through wear or corrosion they should be renewed. If lateral play was evident with the fork in position they should also be renewed. The bushes can be drifted out of position, and new ones tapped in until they seat flush with the outer edge of the fork ends.

10 Check that the pivot shaft is not bent out of true, and clean any corrosion off before reassembly commences. It is advisable to coat the pivot shaft with grease to prevent corrosion taking place. Reassemble the swinging arm fork assembly by reversing the dismantling procedure. Tighten the pivot pin nut to 4.5 kgf m (32 lbf ft).

11 Rear suspension units: examination

1 The swinging arm rear fork assembly is supported by two suspension units, of the hydraulically damped spring type. Each unit consists of a hydraulic damper, effective primarily on rebound, and a concentric spring. It is mounted by way of rubber-bushed lugs at top and bottom.

2 The suspension units are sealed, and therefore no maintenance is feasible. In the event of a unit leaking, or if the damping fails, both units should be renewed as a matched pair.

10.7a Remove and disconnect rear brake light switch ...

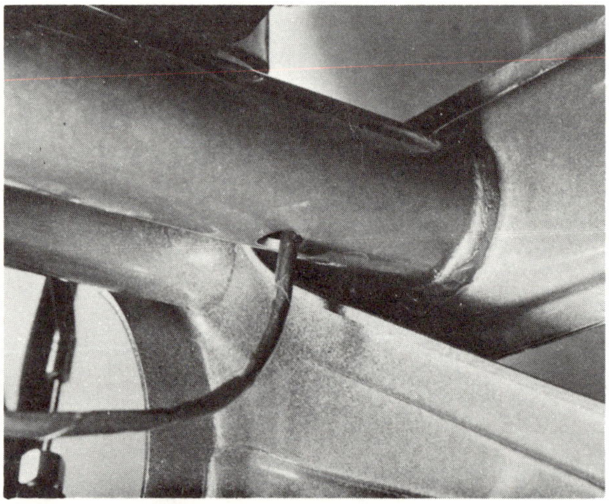

10.7b ... and pull leads clear of swinging arm

10.8a Release pivot shaft nut ...

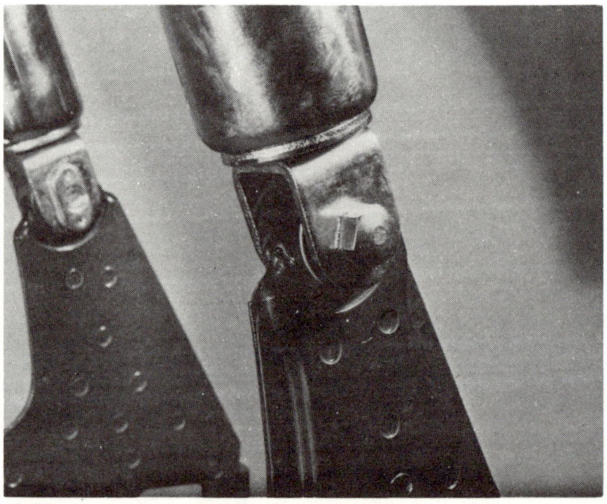

10.8b ... and disconnect suspension units

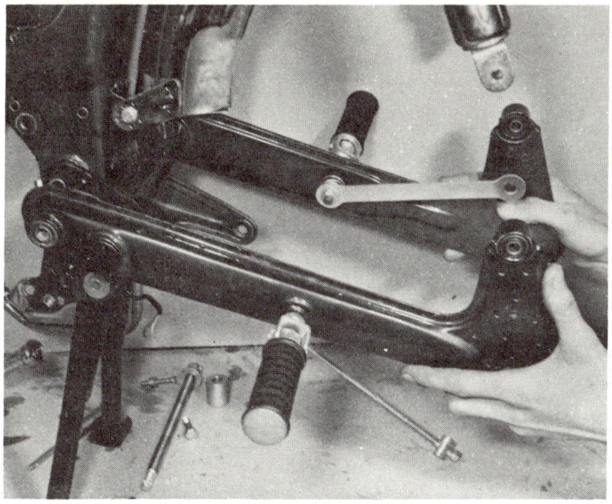
10.8c Swinging arm can now be removed from frame

10.9a Check for wear in swinging arm bushes ...

10.9b ... and at suspension unit mountings

13.1 Propstand and footrests form a combined unit

12 Centre stand: examination

1 The centre stand is mounted on a pivot bolt which passes through the lower part of the frame. A strong return spring, attached to a small lug on the right-hand side of the stand, retracts and holds the stand in position when not in use.
2 Periodically, the pivot pin should be lubricated with grease. This is fairly important as its exposed position renders it susceptible to corrosion if left unattended.
3 Be especially careful to check the condition and correct location of the return spring. If this fails, the stand will fall onto the road in use, and may unseat the rider it it catches in a drain cover or similar obstacle.

13 Prop stand: examination

1 A prop stand is fitted to the left-hand side of the machine for quick parking and for parking on cambered surfaces. It pivots on a bolt passing through the footrest mounting plate, and has a spring which both holds it in the extended position, and also returns it when not in use.
2 As with the centre stand, it is important to check that the stand and its mounting bolt are in sound condition, and that it is kept lubricated. After extended periods of use, the pivot bolt may wear, due to the leverage imposed upon it. If this occurs, it must be renewed before it wears the mounting plate through which it passes.

14 Footrests: examination

1 The footrests comprise an assembly mounted below the frame and can be detached as a complete unit. The assembly is retained by the lower rear engine bolt and one short bolt on the left-hand side. It is easier to remove the silencer, to facilitate removal.
2 Damage is likely only in the event of the machine being dropped. Slight bending can be rectified by stripping the assembly to the bare footrest bar, and straightening the bends by clamping the bar in a vice. A blowlamp should be applied to the affected area to avoid setting up stresses in the material, which may lead to subsequent fracturing.

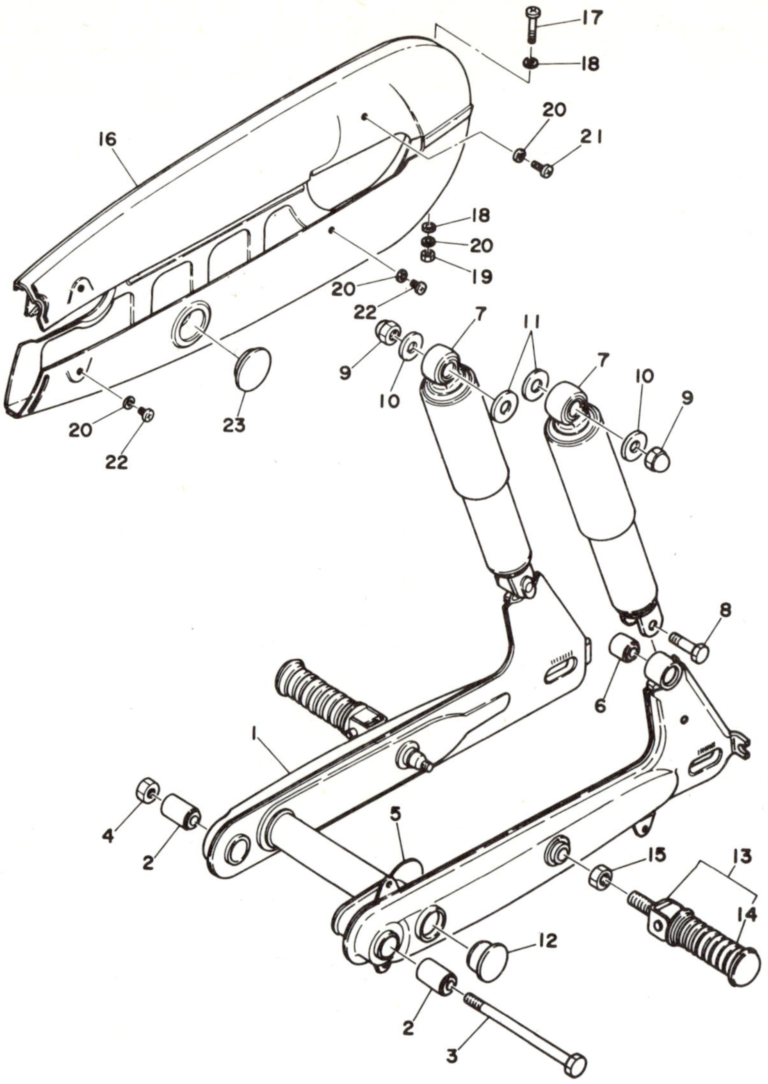

1 *Swinging arm fork*
2 *Bush – 2 off*
3 *Pivot for swinging arm fork*
4 *Nut*
5 *Guard seal*
6 *Rear suspension unit bush – 2 off*
7 *Rear suspension unit – 2 off*
8 *Bolt – 2 off*
9 *Nut – 2 off*
10 *Washer – 2 off*
11 *Washer – 2 off*
12 *Cap*
13 *Pillion footrest assembly – 2 off*
14 *Pillion footrest rubber – 2 off*
15 *Nut – 2 off*
16 *Chain case assembly*
17 *Screw*
18 *Washer – 2 off*
19 *Nut*
20 *Spring washer – 5 off*
21 *Screw*
22 *Screw – 3 off*
23 *Cap*

Fig. 4.4 Swinging arm, rear suspension and chain case

15 Rear brake pedal: examination and renovation

1 The rear brake pedal is mounted on the end of the centre stand pivot shaft. It is retained by an E clip.
2 Should the pedal become bent in an accident, it can be straightened in a similar manner to that given for footrests in the preceding Section. If severely distorted, it should be renewed.

16 Kickstart lever: examination and renovation

1 The kickstart lever is splined and is secured to its shaft by means of a pinch bolt. The kickstart crank swivels so that is can be tucked out of the way when the engine is started. It is held in position on the swivel by a washer and circlip. A spring-loaded ball bearing locates the kickstart arm in either the operating or folded position; if the action becomes sloppy it is probable that the spring behind the ball bearing needs renewing. It is advisable to remove the circlip and washer occasionally, so that the kickstart crank can be detached and the swivel greased.
2 It is unlikely that the kickstart crank will bend in an accident unless the machine is ridden with the kickstart in the operating and not folded position. It should be removed and straightened, using the same technique as that recommended for the footrests.

17 Speedometer head: location and examination

1 The speedometer head is mounted on a bracket on the fork top yoke. On some early models, it is enclosed by a plastic mounting pod, and can be removed after this has been released.
2 The instrument head rarely gives trouble during the normal life of the machine, unless damaged in an accident. Usually, speedometer defects can be traced to the drive cable or gearbox. In the event of speedometer failure, check whether the odometer is still operating. If it is not, the chances are that the cable has broken, and this should be checked first. Should the odometer be operating whilst the speedometer is not, or vice versa, it can be assumed that the instrument is damaged internally.

3 Apart from defects in either the drive or the drive cable, a speedometer that malfunctions is difficult to repair. Fit a new one, or alternatively entrust the repair to a component instrument repair specialist.
4 Remember that a speedometer in correct working order is a statutory requirement in the UK. Apart from this legal requirement, reference to the odometer reading is the best means of keeping in pace with the maintenance schedule.

18 Speedometer drive cable: examination and maintenance

1 It is advisable to detach the cable from time to time in order to check whether it is lubricated adequately, and whether the outer covering is compressed or damaged at any point along its run. Jerky or sluggish movements can often be attributed to a cable fault.
2 As mentioned previously, speedometer malfunctions are often traced to the cable, and a new cable will usually effect a complete cure. Where a cable with a removable inner is fitted, it should be withdrawn and cleaned with a petrol-soaked rag. When regreasing an inner cable, do not lubricate the last six inches of cable where it enters the instrument head. If this precaution is not observed, grease will work into the head and immobilise the movement.

19 Speedometer drive: location and examination

1 The speedometer drive gearbox is an integral part of the front wheel brake plate and is driven internally from the wheel hub. The gearbox rarely gives trouble if it is lubricated whenever the front wheel and brake plate are removed; there is no external grease nipple. If wear in the drive mechanism occurs, the worm complete with shaft can be withdrawn from the brake plate housing after releasing the circlip and unscrewing the threaded body. The drive pinion is retained to the inside of the brake plate by a circlip, in front of the shaped driving plate that takes up the drive from the wheel hub.

20 Cleaning the machine

1 After removing all surface dirt with a rag or sponge washed frequently in clean water, the machine should be allowed to dry thoroughly. Application of car polish or wax to the cycle parts will give a good finish, particularly if the machine has been neglected for a long period.
2 The plated parts of the machine should require only a wipe with a damp rag. If the plated parts are badly corroded, as may occur during the winter when the roads are salted, it is preferable to use one of the proprietary chrome cleaners. These often have an oily base, which will help prevent the corrosion from recurring.
3 If the engine parts are particularly oily, use a cleaning compound such as 'Gunk' or 'Jizer'. Apply the compound whilst the parts are dry and work it in with a brush so that it has the opportunity to penetrate the film of grease and oil. Finish off by washing down liberally with plenty of water, taking care that it does not enter the carburettor or the electrics. If desired, the now clean aluminium alloy parts can be enhanced further by using a special polish such as Solvol 'Autosol', which will fully restore their brilliance.
4 Whenever possible, the machine should be wiped down after is has been used in the wet, so that it is not garaged under damp conditions which will promote rusting. Make sure to wipe the chain and re-oil it, to prevent water from entering the rollers and causing harshness with an accompanying high rate of wear. Remember there is little chance of water entering the control cables and causing stiffness of operation if they are lubricated regularly as recommended in the Routine Maintenance Section.

21 Fault diagnosis: frame and forks

Symptom	Cause	Remedy
Machine veers either to the left or the right with hands off handlebars	Bent frame Twisted forks Wheels out of alignment	Check, and renew. Check, and renew. Check, and re-align.
Machine rolls at low speed	Overtight steering head bearings	Slacken until adjustment is correct.
Machine judders when front brake is applied	Slack steering head bearings	Tighten, until adjustment is correct.
Machine pitches on uneven surfaces	Ineffective fork dampers Ineffective rear suspension units	Check oil content of front forks. Check whether units still have damping action.
Fork action stiff	Fork legs out of alignment (twisted in yokes)	Slacken lower yoke clamps, and fork top bolts. Pump fork several times then retighten from bottom upwards.
Machine wanders. Steering imprecise. Rear wheel tends to hop	Worn swinging arm pivot	Dismantle and renew bushes and pivot shaft.

Chapter 5 Wheels, brakes and tyres

Contents

General description	1
Wheels: examination and renovation	2
Front wheel: removal and replacement	3
Speedometer drive gearbox: examination and maintenance	4
Rear wheel and brake: removal and replacement	5
Brakes: dismantling, examination and replacement	6
Wheel bearings: removal and replacement	7
Adjusting the front brake	8
Adjusting the rear brake	9
Cush drive assembly: examination and replacement	10
Rear wheel sprocket: removal, examination and replacement	11
Final drive chain: examination, adjustment and lubrication	12
Tyres: removal and replacement	13
Tyre valve dust caps	14
Fault diagnosis: wheels, brakes and tyres	15

Specifications

	Front	Rear
Tyres		
Size	2.50 x 18 4PR	2.50 x 18 4PR
Pressure	21 psi (1.5 kg/cm²)	28 psi (2.0 kg/cm²)

Brakes	
Type	Single leading shoe (sls) drum
Drum diameter	110 mm (4.33 in)
Shoe diameter	102 mm (4.02 in)
Shoe width	25 mm (0.98 in)
Spring free length	34.5 mm (1.36 in)
Lining thickness:	
Nominal	4.0 mm (0.16 in)
Minimum	2.0 mm (0.08 in)

1 General description

Both wheels fitted to the Yamaha YB100 models are of the same diameter and width. The front wheel is fitted with a ribbed tyre and a block tread tyre is fitted to the rear as standard. The single leading shoe brakes are of the same diameter and width and utilise identical shoes. The wheels are not interchangeable.

2 Wheels: examination and renovation

1 Place the machine firmly on blocks so that the wheel is raised clear of the ground. Spin the wheel and by using a pointer, such as a screwdriver, check the wheel rim alignment.

Small irregularities can be corrected by tightening the spokes in the affected area, although a certain amount of experience is advisable to prevent over-correction. Any flats in the wheel rim should be evident at the same time. These are more difficult to remove and in most cases it will be necessary to have the wheel rebuilt on a new rim. Apart from the effect on stability a flat will expose the tyre bead and walls to greater risk or damage if the machine is run with a deformed wheel.

2 Check for loose and broken spokes. Tapping the spokes is a good guide to tension. A loose spoke will produce a quite different sound and should be tightened by turning the nipple in an anticlockwise direction. Always recheck for run-out by spinning the wheel again. If the spokes have to be tightened an excessive amount, it is advisable to remove both tyre and tube by following the procedure in Section 13 of this Chapter. This is so that the protruding ends of the spokes can be ground off, to prevent them chafing the inner tube and causing punctures.

3 Front wheel: removal and replacement

1 Place the machine on its centre stand. Disconnect the brake cable at the brake operating lever and release the speedometer drive cable.

2 Remove the split pin and wheel spindle nut and withdraw the wheel spindle.

3 Lift the wheel and brake assembly from the machine. Note the spacer and dust cover fitted on the right-hand side of the hub.

4 To replace the wheel reverse the above procedure. Make sure the brake plate recess locates over the lug on the lower fork leg. Before finally tightening the spindle nut, operate the forks, spin the wheel and operate the brake. Firstly, this aligns

the fork leg and secondly it centralises the brake shoes in the drum. Make doubly certain the brake plate recess is correctly located over the lug on the lower fork leg to prevent the plate from turning when braking. Fit a new split pin in the wheel spindle nut.

4 Speedometer drive gearbox: examination and maintenance

1 The speedometer drive gearbox is incorporated in the brake plate of the front wheel and should give little trouble. Drive from the wheel is via a special driver washer. To obtain access first remove the front wheel as described in the preceding Section and take out the brake plate assembly complete.
2 To dismantle the drive pinion assembly remove the circlip, thrust washer, drive washer, drive pinion and thrust washer. The worm gear is retained in the brake plate by a circlip and threaded body.
3 Assemble in the reverse order of dismantling. Lightly lubricate the gears with high melting point grease. Ensure that the large oil seal is in good condition; renew it if in doubt, since if it is damaged, grease will work through and contaminate the brake linings.

3.1a Slacken adjuster and release trunnion ...

3.1b ... then disengage cable as shown

3.1c Release circlip with pliers ...

3.1d ... and withdraw the speedometer drive cable

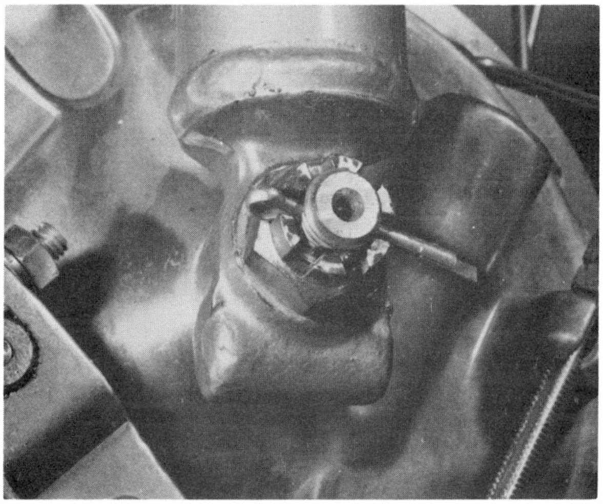
3.2a Remove split pin and spindle nut ...

3.2b ... slacken pinch bolt and withdraw spindle

3.4 Lug must locate in brake plate on reassembly

4.2a Speedometer drive gears reside in brake plate ...

4.2b ... and are driven by lugs on wheel hub

4.2c Threaded boss retains spindle and worm

5 Rear wheel and brake: removal and replacement

1 All models are fitted with a quickly detachable rear wheel which makes it unnecessary to remove the final drive chain.

2 Place the machine on its centre stand. Undo the brake rod adjuster nut, depress the brake pedal and remove the rod from the brake lever trunnion. Do not mislay the trunnion and spring.

3 Remove the 'R' clip and nut on the brake end of the torque arm and free the arm from the backplate.

4 Remove the split pin and undo the rear wheel outer spindle nut. Withdraw the wheel spindle complete with chain adjuster. Note the spacer fitted on the right-hand side. Lift out the brake plate and shoes.

5 Pull the wheel over to the right-hand side to free the cush drive and lift the wheel clear by tilting the machine to the left to obtain clearance.

6 Replacement is by reversing the removal procedure. Do not forget the spacer on the right-hand side of the wheel or the 'R' clip on the torque arm nut. The rear brake will need to be adjusted. Before tightening the wheel spindle nut, spin the wheel and apply the brake to centralise the brake shoes in the drum. Use a new split pin after tightening the spindle nut.

5.2 Release rear brake operating rod ...

5.3 ... and torque arm R-pin and nut

5.4 Wheel spindle can be withdrawn and wheel ...

5.5 ... removed, leaving sprocket hub in place (chain does not need removal

5.6a Wheel and sprocket hub can be removed together

5.6b Do not omit torque arm unit and R-pin

6 Brakes: dismantling, examination and replacement

1 Both front and rear brakes are of the single leading shoe type and for servicing purposes can be treated alike.

2 Remove the wheel and brake as described in Section 3 for the front wheel or Section 5 for the rear wheel. Note that in the case of the rear brake it is not necessary to remove the rear wheel completely, only to proceed as far as withdrawing the wheel spindle and removing the spacer. This then provides sufficient clearance to remove the brake plate and brake shoes.

3 Examine the condition of the brake linings. If they are wearing thin or unevenly the brake shoes should be renewed. The linings are bonded to the brake shoes and cannot be supplied separately.

4 To remove the brake shoes, pull them away from the cam and pivot, and then pull them away from the brake plate in a 'V' formation so that they can be lifted away together with the return springs. When they are well clear of the brake plate, the return springs can be disconnected.

5 Before replacing the brake shoes, check that the operating cam is working smoothly and not binding in its housing. The cam can be removed for greasing by detaching the operating arm from the end of the shaft. The operating arm is located on the cam shaft by splines, and is retained by a pinch bolt, mark both the operating arm and the shaft end before removal to aid correct relocation.

6 Check the inner surface of the brake drum, on which the brake shoes bear. The surface should be free from indentations and score marks, otherwise reduced braking efficiency and accelerated brake lining wear will result. Remove all traces of brake lining dust and wipe the drum surface with a petrol soaked rag, to remove all traces of grease and oil.

7 To reassemble the brake shoes on the brake plate, fit the return springs and pull the shoes apart whilst holding them in the form of a 'V' facing upwards. If they are now located with the brake operating cam and fixed pivot, they can be pushed into position by pressing downwards. Do not use excessive force, or there is risk of distorting the shoes.

6.2 Brake plate assembly can be removed for examination

6.3 Check lining material for wear or contamination

6.4a 'Fold' shoes as shown to remove

6.4b Springs should be fitted as shown here

6.6 Drums should be free from scores or cracks

7 Wheel bearings: removal and replacement

1 The front wheel is supported by two identical journal ball bearings, the rear wheel being supported by two bearings in the wheel hub, and a third contained in the cush drive hub.
2 The bearings are a drive fit in the hub and are removed by driving them out with a drift, working from each side of the hub. When the first bearing emerges from the hub, the hollow distance collar that separates the bearing can be removed.
3 Remove all the grease from the hub and bearings. Check the bearings for play or roughness when they are rotated. If there is any doubt about their condition renew them.
4 Before replacing the bearings, first pack the hub with new high melting point grease, leave sufficient room for expansion of the grease when it becomes hot. Drift the bearings into the hub using a tubular drift, contacting on only the outer ring of the bearing (an appropriate size socket will suffice). Do not forget the distance collar between the bearings.
5 On the rear wheel a third bearing and an additional oil seal is contained in the sprocket carrier/cush drive unit. Drift out the stub spindle and remove the spacer. Prise out the oil seal. The bearing can now be drifted out from the other side of the carrier. Check the bearing as before and again repack with high melting point grease. Refit as for the other wheel bearings. Replace the oil seal (renew if necessary), spacer and spindle.

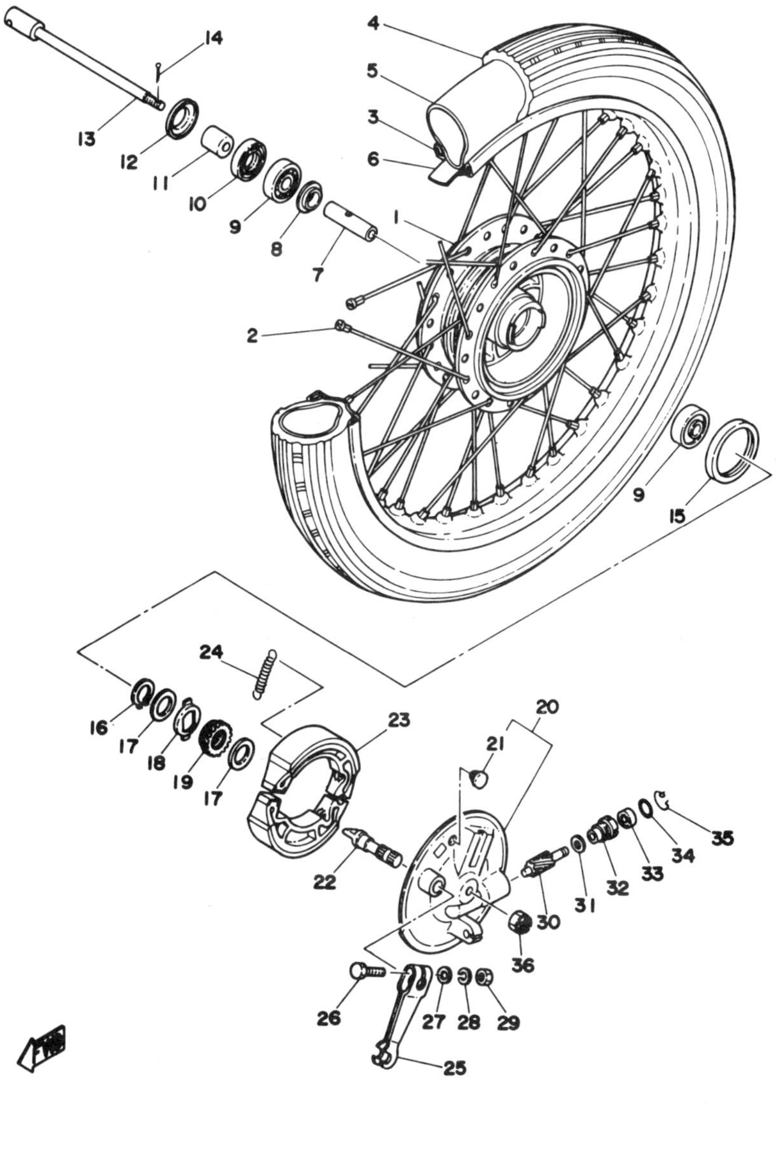

Fig. 5.1 Front wheel

1 Hub
2 Spoke set
3 Rim
4 Front tyre
5 Inner tube
6 Rim tape
7 Bearing spacer
8 Flanged spacer
9 Bearing – 2 off
10 Oil seal
11 Spindle cover
12 Dust cover
13 Wheel spindle
14 Split pin
15 Oil seal
16 Circlip
17 Washer – 2 off
18 Speedometer drive
19 Speedometer drive pinion (27T)
20 Brake plate
21 Grommet
22 Brake operating cam
23 Complete brake shoe – 2 off
24 Brake shoe return spring – 2 off
25 Brake operating arm
26 Bolt
27 Washer
28 Spring washer
29 Nut
30 Speedometer drive take off pinion
31 Washer
32 Bush
33 Oil seal
34 O-ring
35 Stop ring
36 Castellated nut

7.2a Drive out wheel bearings from opposite side

7.2b Spacer has guide washer to aid location

7.4a New bearings should be greased prior to installation

7.4b Use makeshift drift to fit bearings squarely

7.4c Oil seals are inexpensive and should be renewed

7.4d Do not forget this small spacer on front wheel

7.5a This type of bearing has integral seal

7.5b Prise out the oil seal ...

7.5c ... to gain access to circlip

7.5d Do not omit distance piece

7.5e Short tubular spindle fits through bearing

8 Adjusting the front brake

1 The front brake cable adjuster is located at the lower end of the cable, at the front brake plate. Set the cable free play to 5 – 8 mm (0·20 – 0·32 in).

2 Check that the brake pulls off correctly when the handlebar lever is released. Sluggish action is usually indicative of a poorly lubrication control cable, broken or stretched return springs or a tendency for the brake operating arm to bind in its bush. Rubbing brakes affect performance and can cause severe overheating of the lining, drums and wheel bearings.

9 Adjusting the rear brake

1 If the adjustment of the rear brake is correct, the brake pedal will move from 20 – 30 mm (0·8 – 1·2 inches) before the brake begins to operate. Adjustment is made by turning the nut at the end of the brake operating rod to increase or decrease the amount of free play as required.

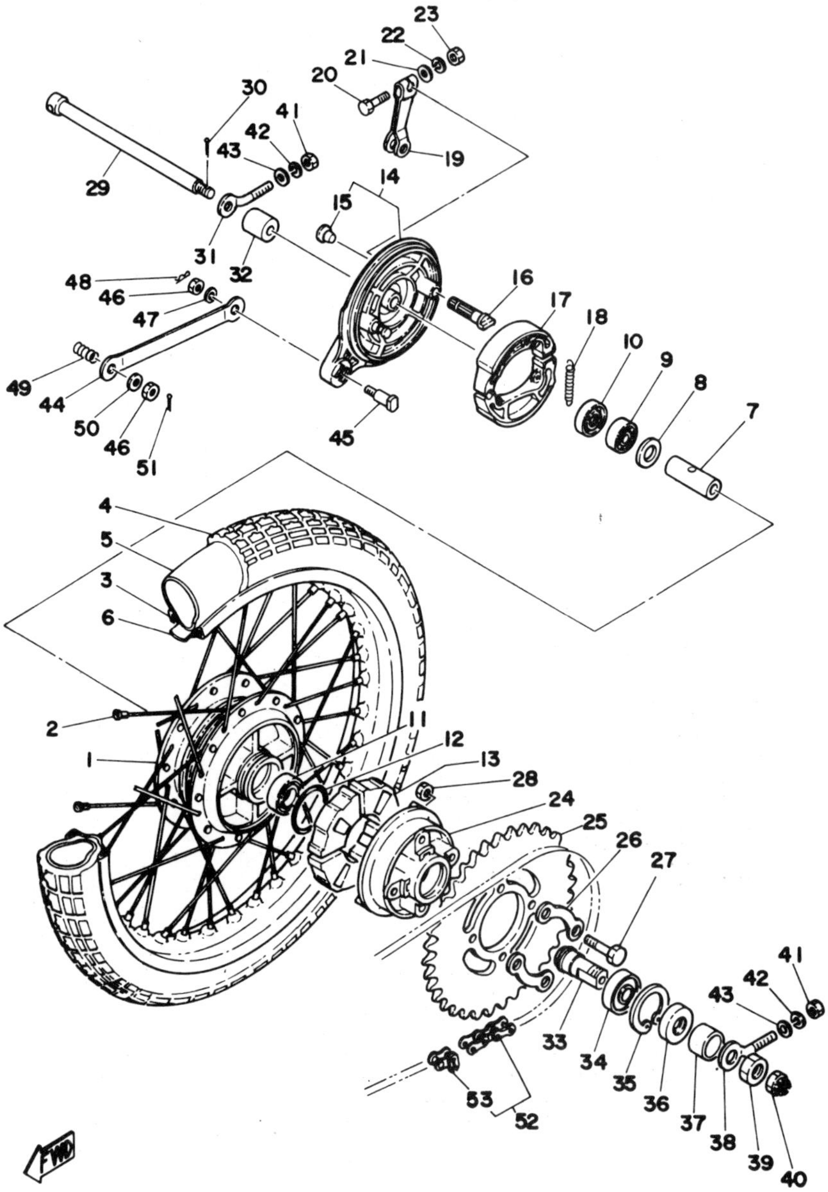

Fig. 5.2 Rear wheel

1 *Hub*
2 *Spoke set*
3 *Rim*
4 *Rear tyre*
5 *Inner tube*
6 *Rim tape*
7 *Bearing spacer*
8 *Flanged spacer*
9 *Bearing*
10 *Oil seal*
11 *Bearing*
12 *O-ring*
13 *Cush drive – 4 off*
14 *Brake plate*
15 *Grommet*
16 *Brake operating cam*
17 *Complete brake shoe – 2 off*
18 *Spring – 2 off*
19 *Brake operating arm*
20 *Bolt*
21 *Washer*
22 *Spring washer*
23 *Nut*
24 *Cush drive plate*
25 *Final drive sprocket*
26 *Lock washer – 2 off*
27 *Bolt – 4 off*
28 *Nut – 4 off*
29 *Wheel spindle*
30 *Split pin*
31 *Right-hand chain adjuster*
32 *Wheel spindle collar*
33 *Sprocket wheel spindle*
34 *Bearing*
35 *Circlip*
36 *Oil seal*
37 *Spindle collar*
38 *Left-hand chain adjuster*
39 *Nut*
40 *Castellated nut*
41 *Nut – 2 off*
42 *Spring washer – 2 off*
43 *Washer – 2 off*
44 *Torque arm*
45 *Bolt*
46 *Nut – 2 off*
47 *Spring washer*
48 *Torque arm clip*
49 *Spring*
50 *Washer*
51 *Split pin*
52 *Final drive chain*
53 *Joining link*

8.1 Adjust cable free play at front wheel

10 Cush drive assembly: examination and replacement

1 The cush drive assembly is contained within the left-hand side of the rear wheel hub. It comprises a synthetic rubber buffer housed within a series of vanes cast in the hub shell. A plate attached to the rear wheel sprocket has six cast-in dogs that engage with slots in these rubbers. The drive to the rear wheel is transmitted through these rubbers, which cushion any surges and roughness in the drive which would otherwise convey the impression of harshness.
2 Under normal riding conditions the cush drive rubbers will continue to be serviceable for an extended length of service. The rubbers should be tested in situ by firmly holding the rear wheel and turning the sprocket alternatively in a clockwise and anti-clockwise direction. If it is evident that the rubbers have become permanently compressed they should be renewed.
3 The cush drive is a push fit in the rear wheel hub and thus presents no problems to replace.

11 Rear wheel sprocket: removal, examination and replacement

1 The rear wheel sprocket can be detached as a separate unit, after the rear wheel has been removed from the frame, as described in Section 5 of this Chapter, and after the removal of the rear drive chain.
2 Remove the nut securing the left-hand rear wheel adjuster, freeing the rear sprocket assembly. The sprocket is retained to the cush drive plate by four nuts and two lockwashers. The tabs on the lockwashers must be knocked down before loosening the nuts.
3 Check the condition of the sprocket teeth. If they are hooked, chipped or badly worn, the sprocket must be renewed. It is considered bad practice to renew one sprocket on its own. The final drive sprockets should always be renewed as a pair and a new chain fitted, otherwise rapid wear will necessitate even earlier renewal on the next occasion.
4 For bearing replacement, refer to Section 7 of this Chapter.

12 Final drive chain: examination, adjustment and lubrication

1 Periodically the tension of the final drive chain will need to be readjusted, to compensate for wear. This is accomplished by slackening the rear wheel nuts after the machine has been placed on the centre stand and drawing the wheel backwards by means of the drawbolt adjusters in the fork ends. The torque arm bolt on the rear brake plate must also be slackened during this operation.
2 The chain is in correct tension if there is from 20 – 25 mm slack. This can be checked via the inspection hole in the chaincase. Always check when the chain is at its tightest point; a chain rarely wears evenly during service.
3 Always adjust the drawbolts an equal amount in order to preserve wheel alignment. The fork ends are marked with a series of horizontal line above the adjusters, to provide a visual check. If desired, wheel alignment can be checked by running a plank of wood parallel to the machine, so that it touches both walls of the rear tyre. If wheel alignment is correct, it should be equidistant from either side of the front wheel tyre, when tested on both sides of the rear wheel. It will not touch the front wheel tyre because this tyre is of smaller cross section. See accompanying diagram.
4 Do not run the chain overtight to compensate for uneven wear. A tight chain will place excessive stresses on the gearbox

11.2 Sprocket is secured by four bolts

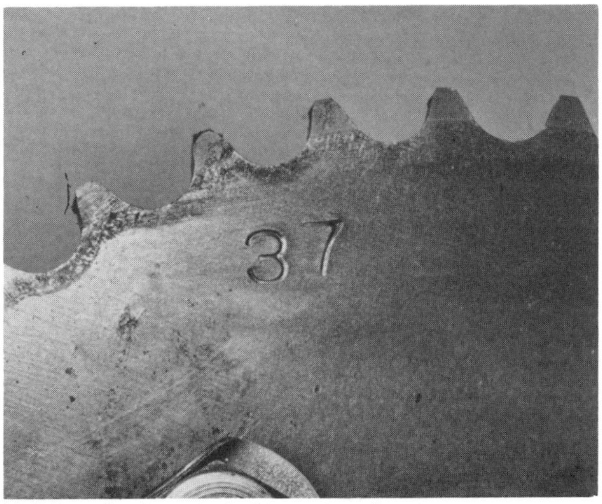

11.3 Sprocket is marked with number of teeth

and rear wheel bearings, leading to their early failure. It will also absorb a surprising amount of power.

5 After a period of running, the chain will require lubrication. Lack of oil will accelerate wear of both chain and sprockets and lead to harsh transmission. The application of engine oil will act as a temporary expedient, but it is preferable to remove the chain and immerse it in a molten lubricant such as 'Linklyfe' or 'Chainguard', after it has been cleaned in a paraffin bath. These latter lubricants achieve better penetration of the chain links and rollers and are less likely to be thrown off when the chain is in motion.

6 To check whether the chain requires replacement, lay it lengthwise in a straight line and compress it endwise until all the play is taken up. Anchor one end and pull on the other in order to take up the end play in the opposite direction. If the chain extends by more than the distance between two adjacent rollers, it should be replaced in conjunction with the sprockets. Note that this check should be made after the chain has been washed out, but before any lubricant is applied, otherwise the lubricant will take up some of the play.

7 When replacing the chain, make sure the spring link is seated correctly, with the closed end facing in the direction of chain travel.

8 The chain fitted is of Japanese manufacture. When renewal is necessary, it should be noted a Renold equivalent, of British manufacture, is available as an alternative. When obtaining a replacement, take along the old chain as a pattern and, if known, a note of the size and number of pitches.

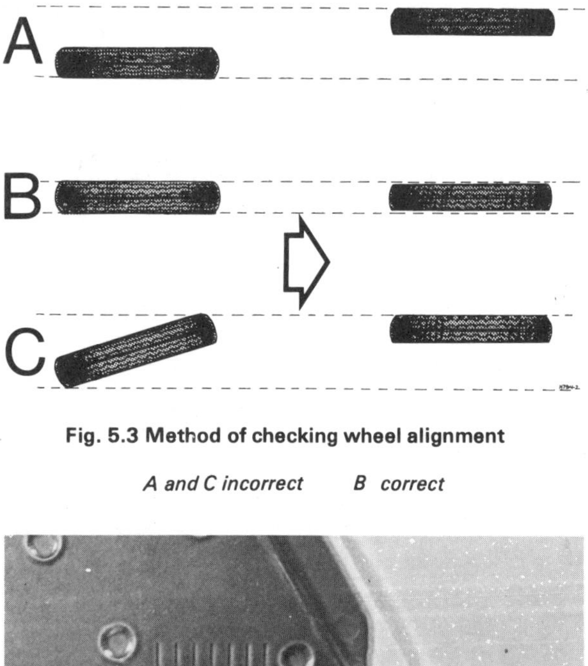

Fig. 5.3 Method of checking wheel alignment

A and C incorrect B correct

13 Tyres: removal and replacement

1 At some time or other the need will arise to remove and replace the tyres, either as the result of a puncture or because replacements are necessary to offset wear. To the inexperienced, tyre changing represents a formidable task yet if a few simple rules are observed and the technique learned, the whole operation is surprisingly simple.

2 To remove the tyre from either wheel, first detach the wheel from the machine by following the procedure in Sectioin 3 or 5 of this Chapter, depending on whether the front or the rear wheel is involved. Deflate the tyre by removing the valve insert and when it is fully deflated, push the bead from the tyre away from the wheel rim on both sides so that the bead enters the centre well of the rim. Remove the locking cap and push the tyre valve into the tyre itself.

3 Insert a tyre lever close to the valve and lever the edges of the tyre over the outside of the wheel rim. Very little force should be necessary; if resistance is encountered it is probably due to the fact that the tyre beads have not entered the well of the wheel rim all the way round the tyre.

4 Once the tyre has been edged over the wheel rim, it is easy to work around the wheel rim so that the tyre is completely free on one side. At this stage, the inner tube can be removed.

5 Working from the other side of the wheel, ease the other edge of the tyre over the outside of the wheel rim which is furthest away. Continue to work around the rim until the tyre is free completely from the rim.

6 If a puncture has necessitated the removal of the tyre, re-inflate the inner tube and immerse it in a bowl of water to trace the source of the leak. Mark its position and deflate the tube. Dry the tube and clean the area around the puncture with a petrol soaked rag. When the surface has dried, apply rubber solution and allow this to dry before removing the backing from the patch and applying the patch to the surface.

7 It is best to use a patch of the self-vulcanising type, which will form a very permanent repair. Note that it may be necessary to remove a protective covering from the top surface of the patch, after it has sealed in position. Inner tubes made from synthetic rubber may require a special type of patch and adhesive, if a satisfactory bond is to be achieved.

8 Before replacing the tyre, check the inside to make sure the agent which caused the puncture is not trapped. Check the outside of the tyre, particularly the tread area to make sure nothing is trapped which may casue a further puncture.

9 If the inner tube has been patched on a number of past occasions, or if there is a tear or large hole, it is preferable to discard it and fit a replacement. Sudden deflation may cause an accident, particularly if it occurs with the front wheel.

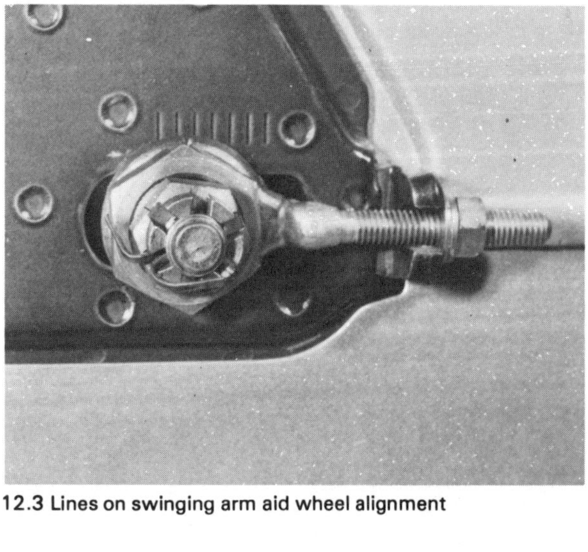

12.3 Lines on swinging arm aid wheel alignment

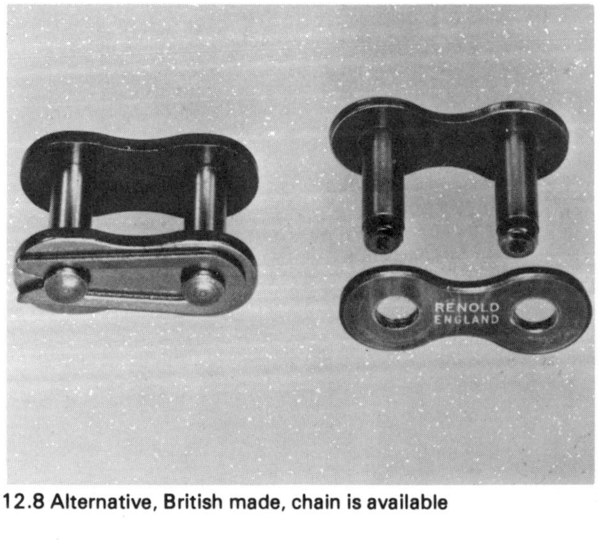

12.8 Alternative, British made, chain is available

Tyre removal: Deflate inner tube and insert lever in close proximity to tyre valve

Use two levers to work bead over the edge of rim

When first bead is clear, remove tyre as shown

Tyre fitting: Inflate inner tube and insert in tyre

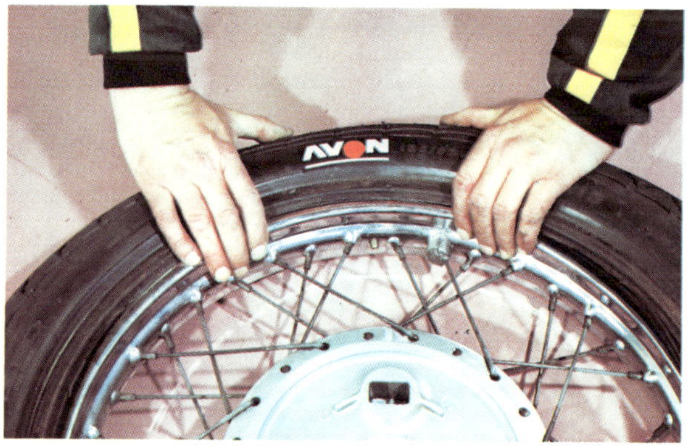

Lay tyre on rim and feed valve through hole in rim

Work first bead over rim, using lever in final section

Use similar technique for second bead, finish at tyre valve position

Push valve and tube up into tyre when fitting final section, to avoid trapping

10 To replace the tyre, inflate the inner tube sufficiently for it to assume a circular shape but only just. Then push it into the tyre so that it is enclosed completely. Lay the tyre on the wheel at an angle and insert the valve through the rim tape and the hole in the wheel rim. Attach the locking cap on the first few threads, sufficient to hold the valve captive in its correct location.
11 Starting at the point furthest from the valve, push the tyre bead over the edge of the wheel rim until it is located in the central well. Continue to work around the tyre in this fashion until the whole of one side of the tyre is on the rim. It may be necessary to use a tyre lever during the final stages.
12 Make sure there is no pull on the tyre valve and again commencing with the area furthest from the valve, ease the other bead of the tyre over the edge of the rim. Finish with the area close to the valve, pushing the valve up into the tyre until the locking cap touches the rim. This will ensure the inner tube is not trapped when the last section of the bead is edged over the rim with a tyre lever.
13 Check that the inner tube is not trapped at any point. Re-inflate the inner tube and check that the tyre is seating correctly around the wheel rim. There should be a thin rub moulded around the wall of the tyre on both sides, which should be equidistant from the wheel rim at all points. If the tyre is unevenly located on the rim, try bouncing the wheel when the tyre is at the recommended pressure. It is probable that one of the beads has not pulled clear of the centre well.
14 Always run the tyres at the recommended pressures, and never under or over-inflate. The correct pressures for solo use are given in the Specifications Section of this Chapter. If a pillion passenger is carried, increase the rear tyre pressure as recommended.
15 Tyre replacement is aided by dusting the side walls, particularly in the vicinity of the beads, with a liberal coating of French chalk. Washing-up liquid can also be used to good effect, but this has the disadvantage of causing the inner surfaces of the wheel rim to rust.
16 Never replace the inner tube and tyre without the rim tape in position. If this precaution is overlooked there is a good chance of the spoke nipples chafing the inner tube causing punctures.
17 Never fit a tyre which has a damaged tread or side walls. Apart from the legal aspects, there is a very great risk of blow-out, which can have serious consequences on any two wheel vehicle.
18 Tyre valves rarely give trouble, but it is always advisable to check whether the valve itself is leaking before removing the tyre. Do not forget to fit the dust cap which forms an effective second seal.

14 Tyre valve dust caps

1 Tyre valve dust caps are often left off when a tyre has been replaced, despite the fact that they serve an important two-fold function. Firstly, they prevent dirt or other foreign matter from entering the valve and causing the valve to stick open when the tyre pump is next applied. Secondly, they form an effective second seal so that in the event of the tyre valve leaking air will not be lost.

15 Fault diagnosis: wheels, brakes and tyres

Symptom	Cause	Remedy
Handlebars oscillate at low speeds	Buckled front wheel Incorrectly fitted front tyre	Remove wheel for specialist attention. Check whether line around bead is equidistant from rim.
Forks 'hammer' at high speeds	Front wheel out of balance	Add weights until wheel will stop in any position.
Brakes grab, locking wheel	Ends of brake shoes not chamfered	Remove brake shoes and chamfer ends.
Brakes feel spongy	Stretched brake operating cables, weak pull-off springs	Replace cables and/or springs, after inspection.
Tyres wear more rapidly in middle of tread	Over-inflation	Check pressures and run at recommended settings.
Tyres wear rapidly at outer edge of tread	Under-inflation	Ditto.

Chapter 6 Electrical system

Contents

General description	1	Stop and tail lamp: replacing the bulb	9
Charging system testing: general	2	Flashing indicator lamps	10
Charging system: checking the output	3	Flasher relay: location and renewal	11
Rectifier: location and testing	4	Speedometer and warning lamps: bulb renewal	12
Battery: charging procedure and maintenance	5	Brake light switches: location and renewal	13
Fuse: location and replacement	6	Horn: location and examination	14
Headlamp: replacing bulbs and adjusting beam height	7	Wiring: layout and examination	15
Switches: location and maintenance	8	Fault diagnosis: electrical system	16

Specifications

Battery

Make	GS	
Type	6N4–2A–2 or 6N4A–4D	
Voltage	6 volt	
Capacity	4 Ah	
Charge rate	0.4 amps for 10 hours	
Specific gravity	1.260 @ 20° C	

Flywheel generator

Type	F2 N3	
Voltage	6 volt	
	YB100	**YB100E**
Charging output	0.7 A ± 0.3 A @ 3000 rpm	0.1A or more @ 2500 rpm
	4.7 A or less @ 8000 rpm	2.3 A ± 0.5 A @ 8000 rpm
Charging coil resistance	Green lead to earth 0.17 ohms ± 10% @ 20° C	Green/red lead to earth 0.38 ohms ± 10% @ 20° C
Lighting output	6.2V or more @ 2500 rpm	5.7V or more @ 2500 rpm
	8.5V or less @ 8000 rpm	8.0V or less @ 8000 rpm
Lighting coil resistance	Yellow lead to earth 0.52 ohms ± 10% @ 20° C	Yellow lead to earth 0.28 ohms ± 10% @ 20° C

Rectifier

Type	Silicon diode, half-wave
Make	Stanley
Model	DE 4504
Capacity	4 amps

	YB100	YB100E
***Bulbs**		
Headlamp	6V 25/25W	6V 30/30W
Stop/tail lamp	6V 5/21W	6V 5.3/25W
Indicator lamp	6V 10W	6V 17W
Speedometer lamp	6V 3W	6V 3W
Pilot lamp	6V 3W	6V 3W
Horn		
Make	Nikko	Nikko
Model	YF–6	MF–6
Indicator relay		
Make	Mitsuba	Nippon Denso
Type	FR–6	FN637L
Fuse		
Rating	10 amps	10 amps

* Note: Bulb ratings may vary between models and in different countries. When buying replacements take the old bulb as a pattern

1 General description

The Yamaha YB100 models are equipped with a simple electrical system, powered by a flywheel generator assembly mounted on the left-hand end of the crankshaft. The generator is in fact a small alternator, having two coils, one of which is used as an ignition source coil. The remaining coil provides power for the electrical system and for battery charging.

The alternating current (ac) output of the generator must be converted to direct current (dc) before it can be fed to the electrical system. This is achieved by way of a silicon rectifier, which effectively blocks half of the output sine wave by acting as an electronic switch. For obvious reasons, the system is known as half-wave rectification.

The electrical system is protected by a single 10 amp fuse carried in line in the positive (+) battery lead.

2 Charging system testing: general

1 To carry out the various tests on the charging system, it will be necessary to use a pocket multimeter. These are inexpensive test instruments, and are well worth investing in. For the purposes of the tasks described in this Chapter, it has been assumed that the owner has a suitable meter in his or her possession, and is familiar with its use.

2 Although the tests themselves are quite straightforward, there is some danger of damaging certain components if wrong connections are made. It is recommended, therefore, that if the owner feels unsure about the test procedures, that they be left to a qualified Yamaha Service Agent who will have in his possession the necessary expertise and equipment to effect an economical repair.

3 It will be noted that the ignition source coil is not dealt with in this Chapter. This is covered in Chapter 3, along with the rest of the ignition system.

3 Charging system: checking the output

1 The charging system is normally a reliable arrangement, giving long service and requiring little or no attention. If electrical problems appear, try to establish the exact nature of the problem, eg; persistent over or under-charging. It is essential that the battery is in good condition and fully charged, and this should always be checked first. Next make a full check of the wiring and connections.

2 The charging current can be measured by connecting an ammeter between the positive (+) battery lead and the positive battery terminal. Start the engine, and note the reading obtained at various speeds. If the reading shown does not correspond with the figures shown in the Specifications Section, the charging coil resistance should be measured.

3 Trace the generator output leads back to their connectors. On early models separate connectors are used, and can be found inside the frame after the left-hand side cover and battery holder have been released. On later machines, a combined connector is plugged into the plastic battery holder.

4 Using a multimeter set on the resistance, or ohms, scale, measure the resistance of the charging coil by connecting one probe to the green (YB100) or green/red (YB100E) lead, and the remaining probe to earth. If the reading obtained is significantly different from that given in the Specifications, or if the coil is completely open or short circuited it may be assumed that the coil is defective and requires renewal.

5 To measure the lighting coil voltage, set the multimeter to the 20V ac setting and connect the positive (+) probe to the yellow output lead and the negative (–) probe to earth. Start the engine, and check the voltage reading at 2500 rpm and 8000 rpm, comparing the readings obtained with those given in the Specifications, noting that the lights should be switched on during the test. A number of factors can affect this reading, namely, broken wiring, bad connections, burnt out bulbs or bulbs of the wrong value. These areas should be checked and rectified before proceeding further.

6 With the engine and lights switched off, leave the multimeter connections as described in paragraph 5, set the meter on resistance, or ohms, and measure the lighting coil resistance. As with the charging coil resistance described earlier, the resistance figures should agree closely with those given in the Specifications.

7 It should be noted that a fault in the charging coil will affect the lighting coil and vice versa, as the two share a common source. It is recommended, therefore, that a thorough check of the entire system is made, to make fault finding more accurate.

4 Rectifier: location and testing

1 The rectifier is mounted beneath the left-hand side cover, adjacent to the battery. If the rectifier is suspected of malfunctioning it can be checked very easily using a multimeter set on the resistance scale. Connect the meter's red probe (+) to the red terminal of the rectifier, and the black probe (–) to the white rectifier terminal. For the purposes of this test, the rectifier leads should be detached. With the test probes connected as described above, a low resistance should be indicated. Reversing the probes should indicate no continuity. If this is not the case, the rectifier can be assumed to be defective and should be renewed.

3.4 Lighting/charging coil is incorporated in generator

4.1 Rectifier is mounted on battery bracket (removed for clarity)

5 Battery: charging procedure and maintenance

1 The normal charging rate for the 4 amp hour battery is 0.4 amps. A more rapid charge, not exceeding 1 amp can be given in an emergency. The higher charge rate should, if possible, be avoided since it will shorten the working life of the battery.

2 Make sure that the battery charger connections are correct, red to positive and black to negative. It is preferable to remove the battery from the machine whilst it is being charged and to remove the vent plug from each cell. When the battery is reconnected to the machine, the black lead must be connected to the negative terminal and the red lead to the positive. This is most important, as the machine has a negative earth system. If the terminals are inadvertently reversed, the electrical system will be damaged permanently. The rectifier will be destroyed by a reversal of the current flow.

3 The electrolyte level of the battery should be maintained between the upper and lower limits marked on the case by topping up with distilled water (unless spillage has occurred when it should be topped up with acid of the correct specific gravity). If, when the battery is in a fully charged condition, (corresponding to approximately 6.6 volts) the specific gravity lies much below 1.26 – 1.28 at 20°C, the electrolyte should be replaced by fresh sulphuric acid of the correct specific gravity (1.26 - 1.28 at 20°C).

4 If the machine will not be used for a while, to prevent deterioration, the battery should be recharged every six weeks or so. If the battery is left in a discharged condition for any length of time the plates will sulphate and render it inoperative.

5 If the battery case is cracked or leaking, a replacement battery should be obtained, since it is not often that an effective repair can be made. A leaking battery should never be used, since the acid will severely corrode the cycle parts. If any acid is spilt over the machine or rider, it should be washed off immediately with plenty of water.

6 Fuse: location and replacement

1 The main fuse is fitted in line in the battery positive lead. It is located in a white plastic fuseholder fitted beneath the left-hand side cover, and is rated at 10 amps.

2 Before replacing a fuse that has blown, check that no obvious short circuit has occurred, otherwise the replacement fuse will blow immediately it is inserted. It is always wise to check the electrical circuit thoroughly to trace the fault and eliminate it.

3 When a fuse blows while the machine is running and no spare is available, a 'get you home' remedy is to remove the blown fuse and wrap it in silver paper before replacing it in the fuseholder. The silver paper will restore the electrical continuity by bridging the broken fuse wire. This expedient should **never** be used if there is evidence of a short circuit or other major electrical fault, otherwise more serious damage will be caused. Replace the 'doctored' fuse at the earliest possible opportunity, to restore full circuit protection.

7 Headlamp: replacing bulbs and adjusting beam light

1 To remove the headlamp rim, detach the small screw on the right-hand underside of the headlamp shell. The rim can then be prised off, complete with the reflector unit.

2 The main bulb is a twin filament type, to give a dipped beam facility. The bulb holder is attached to the back of the reflector by a rubber sleeve, which fits around a flange in the reflector and the flange of the bulb holder. An indentation in the bulb holder orifice and a projection on the bulb holder end ensures the bulb is always replaced in the same position so that the focus is unaltered.

3 It is not necessary to re-focus the headlamp when a new bulb is fitted. Apart from the just-mentioned method of location, the bulbs used are of the pre-focus type, built to a precise specification. To release the bulb holder, twist and lift away.

4 The pilot lamp bulb, like the bulbholder has a bayonet fitting. It is protected by a rubber sleeve. Remove the bulb holder first, then the bulb.

5 The main headlamp bulb is rated at 25/25W, 6 volts and the pilot lamp bulb at 4W, 6 volts. Variations in the wattage may occur according to the country or state for which the machine is supplied.

6 Beam alignment is adjusted by tilting the headlamp after the two retaining bolts have been slackened and then retightening them after the correct beam height is obtained, without moving the setting.

7 To obtain the correct beam height, place the machine on level ground facing a wall 25 feet distant, with the rider seated normally. The height of the beam centre should be equal to the height of the centre of the headlamp from the ground, when the dip switch is in the main beam position. Furthermore, the concentrated area of light should be centrally disposed. Adjustments in either direction are made by rearranging the angle of the headlamp, as described in the preceding paragraph. Note that a different beam setting will be needed when a pillion passenger is carried. If a pillion passenger is carried regularly, the passenger should be seated in addition to the rider when the beam setting adjustment is made.

8 The above instructions for beam setting relate to the requirements of the United Kingdom's transport lighting regulations. Other settings may be required in countries other than the UK.

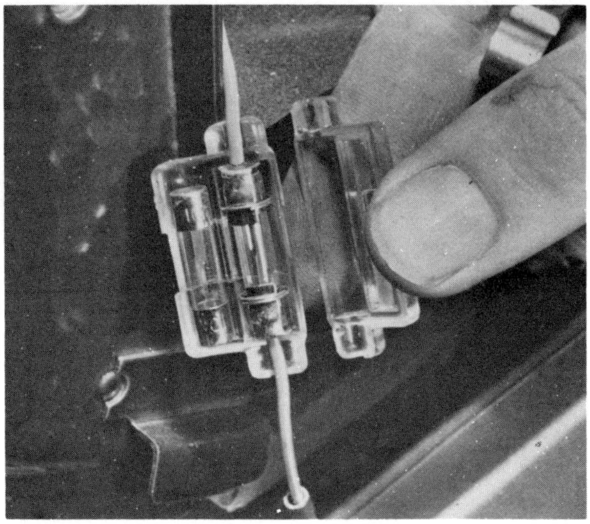

6.1 Fuse holder contains spare fuse

7.1 Slacken rim screw, and displace light unit

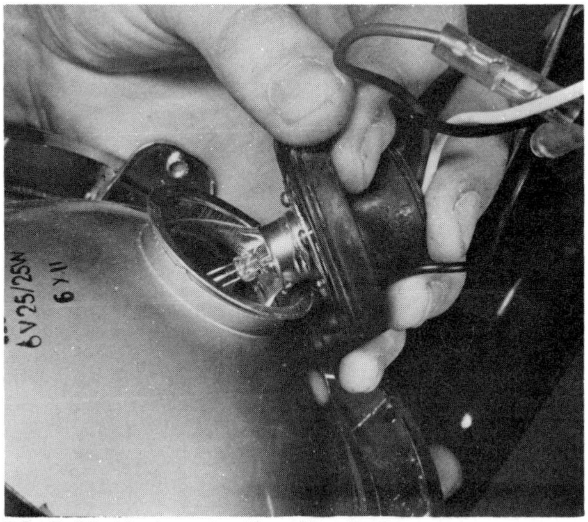

7.2a Bulbholder is retained by rubber flange

7.2b Bulb is retained by headed pins as shown

7.4 Pilot bulb is a bayonet fitting in reflector

7.6 Slacken indicator stalks/headlamp bolts for adjustment

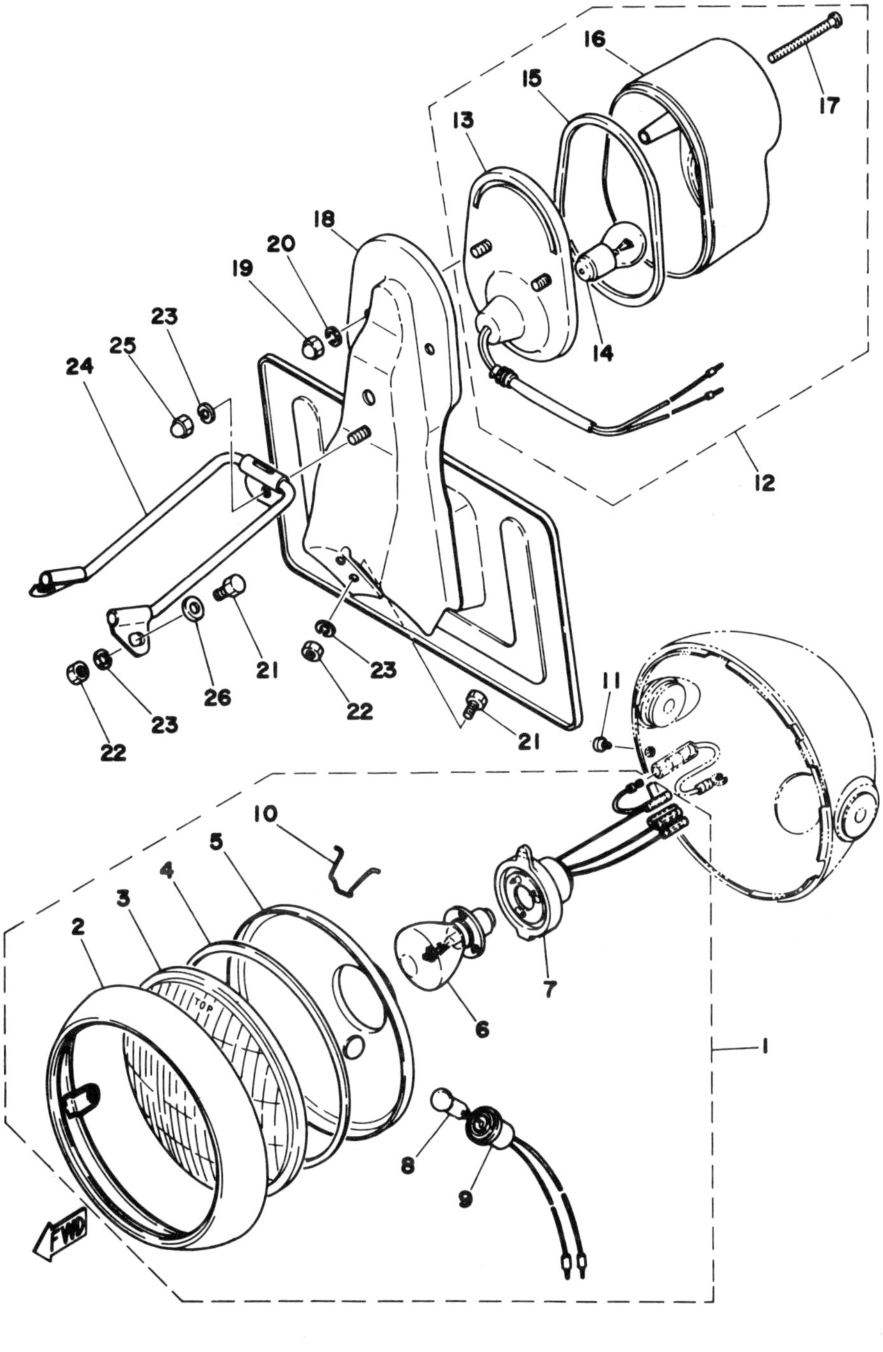

Fig. 6.1 Headlamp and Tail lamp

1 Complete headlamp unit
2 Rim
3 Lens
4 Sealing gasket
5 Reflector
6 Headlamp main bulb
7 Bulb holder
8 Pilot bulb (UK model)
9 Bulb holder and lead wire (UK model)
10 Reflector retaining clip – 4 off
11 Rim securing screw
12 Complete tail lamp unit
13 Backing plate
14 Bulb
15 Sealing gasket
16 Tail lamp lens
17 Screw – 2 off
18 Rear lamp and number plate bracket
19 Nut – 2 off
20 Spring washer – 2 off
21 Bolt – 6 off
22 Nut – 6 off
23 Spring washer – 7 off
24 Number plate bracket stay
25 Nut
26 Washer – 2 off

8 Switches: location and maintenance

1 The main switch is located near the centre of the handlebar assembly and is mounted on the top yoke. It is a key operated device, controlling the ignition and lighting circuits. In the event of a malfunction, it will be necessary to fit a replacement unit. The switch must be considered a sealed unit, and as such any attempt at repair is impractical.

2 The dipswitch and horn button are incorporated in the left-hand handlebar switch unit, the indicator switch being located on the right-hand side. Each unit is enclosed in a light-alloy housing which can be separated after releasing the screws from the underside of the unit.

3 In the event of failure of any of these switches, the switch assembly must be replaced as a complete unit since it is not practicable to effect a permanent repair. Before condemning the unit, however, check that the malfunction is not due to dirty or burnt contacts. It may be possible to restore these with a switch cleaning spray or by burnishing them with fine wet or dry paper.

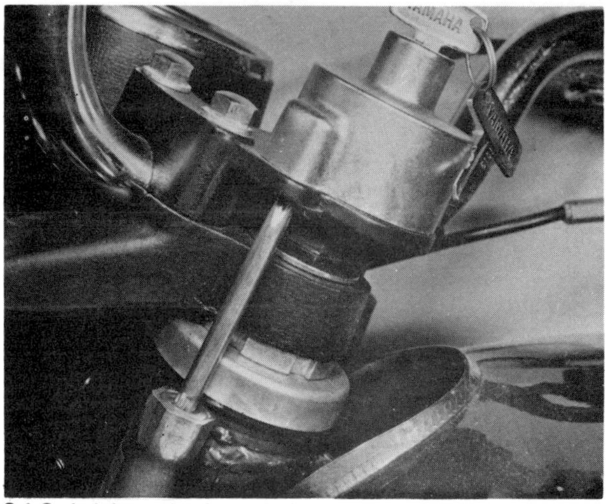

8.1 Switch is retained to yoke by screws

8.2a Left-hand switch controls dip and horn functions ...

8.2b ... whilst right-hand unit operates indicators

8.3a Switch can be removed in two halves ...

8.3b ... and contacts cleaned with aerosol spray

9 Stop and tail lamp: replacing the bulb

1 The tail lamp is fitted with a twin filament bulb to illuminate the rear number plate and rear of the machine, and to give visual warning when the rear brake is applied. To gain access to the bulb remove the plastic lens cover, which is retained by two long screws. Check that the gasket between the lens cover and the main body of the lamp is in good condition.
2 The bulb has a bayonet fitting and has staggered pins to prevent the bulb contacts from being reversed.
3 If the tail lamp bulb keeps blowing, suspect either vibration of the rear mudguard or more probably an intermittent earth connection.

10 Flashing indicator lamps

1 The forward facing indicator lamps are connected to 'stalks' which are attached to the headlamp brackets. The hollow stalks carry the leads to the lens unit. The rear facing lamps are mounted on similar short stalks, at a point immediately to the rear of the dualseat.
2 In each case, access to the bulb is gained by removing the plastic lens cover, which is retained by two screws. Bayonet fitting bulbs of the single filament type are used.

11 Flasher relay: location and renewal

1 The flasher relay is retained by a rubber mounting behind the left-hand side cover. The unit is designed to switch the indicator lamps on and off at a rate of 85 times per minute.
2 Most indicator failures are attributable to damaged wiring, dirty connections or blown bulbs, and these areas should be checked first. For example, if the rate of flash on one side of the machine is noticeably slower than the other, it will almost invariably be caused by a poor connection or by bulbs of the wrong rating being used. The bulb wattages must match the rating of the flasher relay, and for this reason, always check the wattage of replacement bulbs carefully.
3 If a fault is traced to the unit itself, there is no option other than renewal. The relay is sealed, and cannot be repaired, if defective. Great care should be taken when handling the flasher relay, as it can easily be damaged if dropped.

12 Speedometer and warning lamps: bulb renewal

1 Various instrument panel arrangements have been used on the YB100 range, each differing in its method of fitting. On early models where the warning lamps are incorporated in the speedometer head, release the drive cable at the knurled ring, remove the two R-pins and special washers, and lift the instrument head away from its bracket. The rubber bulbholders are a push fit in the base, the bulbs being a bayonet fitting.
2 A second arrangement employs a moulded plastic pod containing the speedometer head and separate warning lamps. To gain access it is necessary to release the speedometer drive cable and the two nuts which secure the speedometer to the base of the pod (see photographs). Ensure the various grommets are not lost during dismantling.
3 The latest type of warning lamp cluster is integral with the ignition switch shroud. Access to the bulbholder block is gained after releasing the two screws which retain the assembly to its mounting bracket. In each case, check that any replacement bulb is of the same value as the defective one. It is advisable to take the blown bulb as a pattern when purchasing replacements.

13 Brake light switches: location and renewal

1 Most of the YB100 models make use of two independent switches to operate the brake light at the rear of the machine. The front brake light is fitted to late models, and is mounted on the underside of the front brake lever assembly. The switch operates automatically when the front brake is applied, and requires no adjustment. Early machines were not fitted with a front brake light switch.
2 The rear brake light switch is mounted in a recess in the swinging arm fork, to which it is retained by a single screw. The switch is operated by way of a small spring connecting its operating arm to the brake pedal. Like the front switch, no provision is made to allow for adjustment, as the movement of the pedal controls its action, and this is unaffected by brake adjustment. If either of the switches fail, it will be necessary to renew them.

9.1 Rearlamp lens is secured by two screws

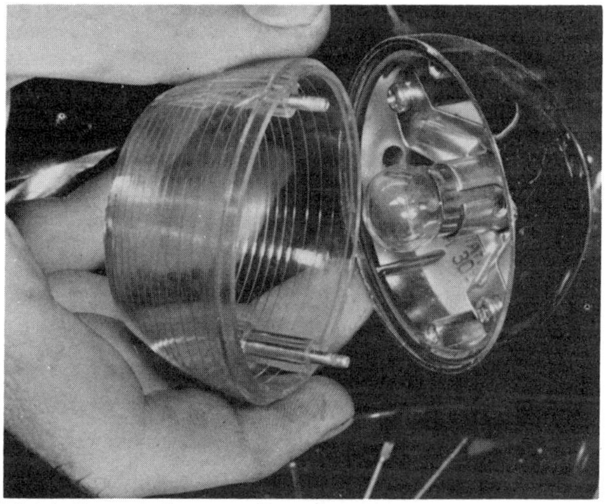
10.1 Indicator bulbs are easily renewed

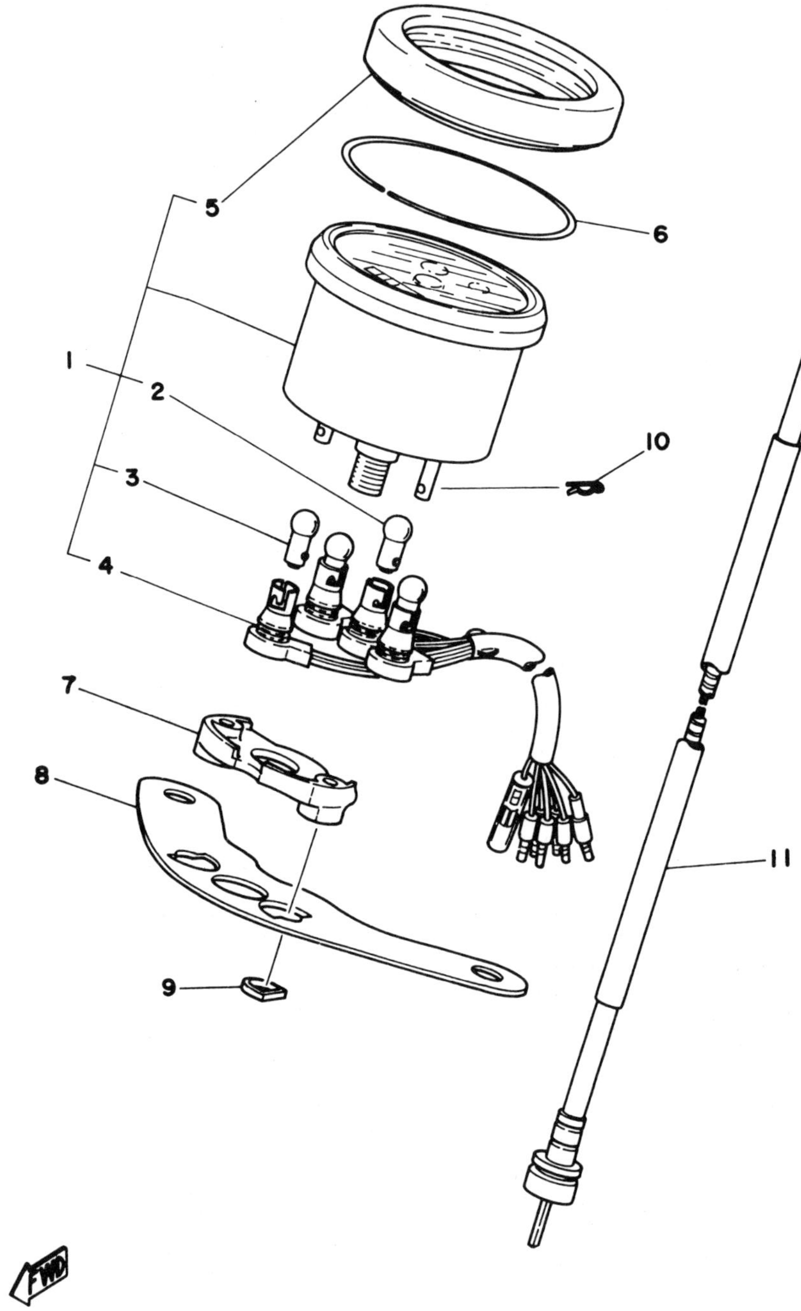

Fig. 6.2 Speedometer

1 *Speedometer assembly*
2 *Bulb*
3 *Bulb*
4 *Bulb holder*
5 *Rubber surround*
6 *Sealing ring*
7 *Vibration insulator*
8 *Mounting bracket*
9 *Washer – 2 off*
10 *Mounting clip – 2 off*
11 *Speedometer drive cable assembly*

12.1a On early models, warning lamps are housed in pod

12.1b Pod halves secured by two nuts (arrowed). Release cable

12.1c Remove the nuts, spring and plain washers ...

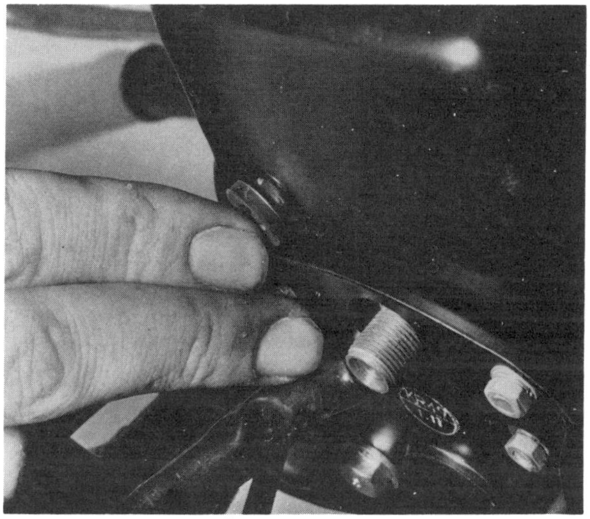
12.1d ... followed by rubber grommets

12.1e Unit can be separated for access to bulbs

13.2a Rear brake switch is mounted in swinging arm

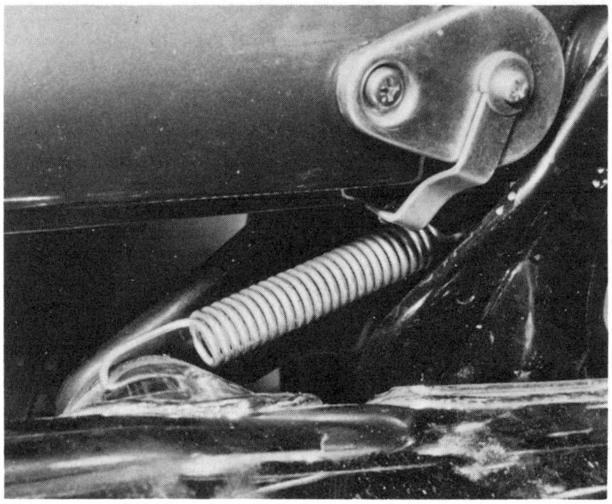
13.2b ... and is operated by spring from brake pedal

14 Horn: location and examination

1 The horn is of the conventional electromagnetic type, and is mounted beneath the petrol tank. The unit will normally function for extended periods, requiring no maintenance. If a fault develops, check that the various connections are secure and that the horn switch contacts are in good condition.

2 If this fails to restore the horn to working order, some adjustment is possible by way of an adjustment screw and locknut. The screw should be turned *gently* by a fraction of a turn at a time until the horn is restored to working order. Do not omit to re-tighten the locknut after the adjustment has been made.

15 Wiring: layout and examination

1 The wiring harness is colour-coded and will correspond with the accompanying wiring diagram. Where socket connectors are used, they are designed so that reconnection can be made in the correct position only.

2 Visual inspection will show whether there are any breaks or frayed outer coverings which will give rise to short circuits. Another source of trouble may be the snap connectors and sockets, where the connector has not been pushed fully home in the outer housing.

3 Intermittent short circuits can often be traced to a chafed wire that passes through or is close to a metal component such as a frame member. Avoid tight bends in the lead or situations where a lead can become trapped between casings.

14.1 Horn is mounted beneath petrol tank

15.1 Wiring is colour-coded to aid fault-finding

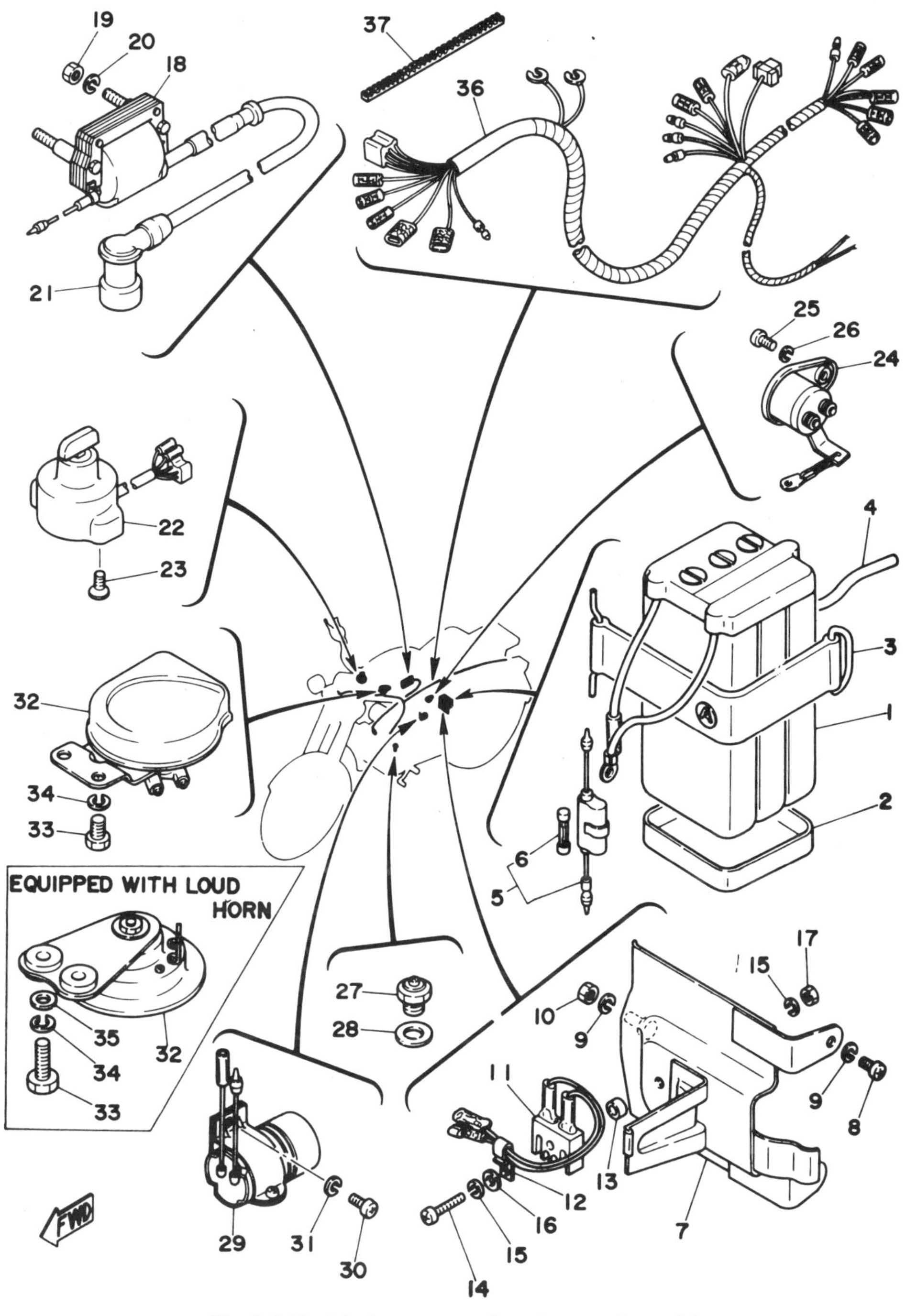

Fig. 6.3 Electrical component location – early models

1 Battery
2 Battery carrier
3 Battery clamp
4 Vent pipe
5 Fuse holder assembly
6 Fuse – 2 off
7 Rectifier mounting plate
8 Screw – 2 off
9 Spring washer– 3 off
10 Nut
11 Rectifier
12 Wire holder
13 Rectifier collar
14 Screw
15 Spring washer – 2 off
16 Washer
17 Nut
18 Ignition coil assembly
19 Nut – 2 off
20 Spring washer – 2 off
21 Plug cap
22 Ignition switch assembly
23 Screw – 2 off
24 Stop lamp switch
25 Screw
26 Spring washer
27 Neutral switch
28 Sealing washer
29 Flasher relay unit
30 Screw
31 Spring washer
32 Horn
33 Bolt – 2 off
34 Spring washer – 2 off
35 Washer – 2 off
36 Wiring harness
37 Cable grommet

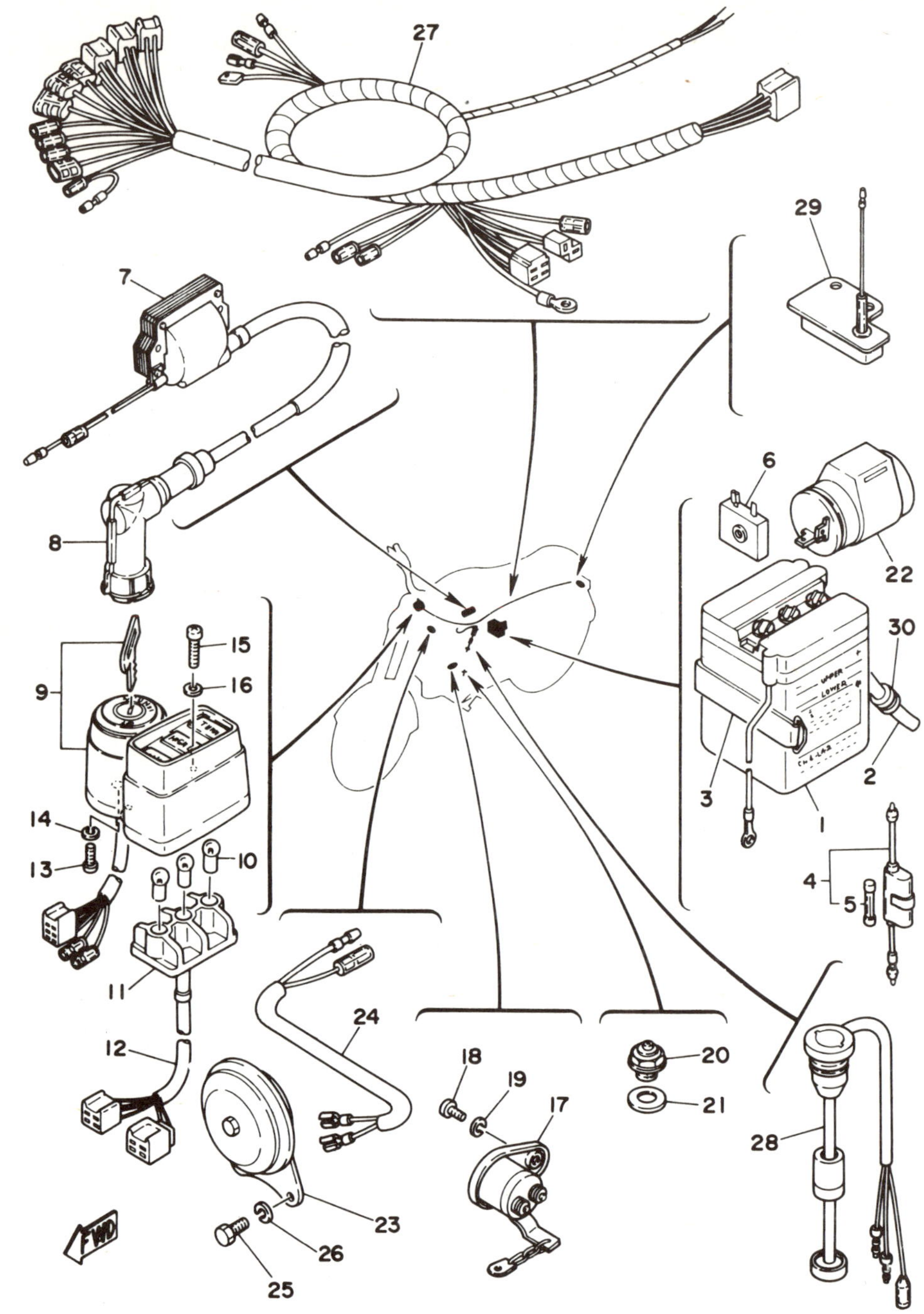

Fig. 6.4 Electrical component location – late models

1	*Battery*	*11*	*Warning lamp bulb holder assembly*	*21*	*Sealing washer*
2	*Battery breather pipe*	*12*	*Warning lamp cable assembly*	*22*	*Flasher unit*
3	*Battery retaining strap*	*13*	*Screw – 2 off*	*23*	*Horn*
4	*Fuse holder*	*14*	*Spring washer – 2 off*	*24*	*Horn cable assembly*
5	*Fuse – 2 off*	*15*	*Screw – 2 off*	*25*	*Bolt – 2 off*
6	*Rectifier assembly*	*16*	*Washer – 2 off*	*26*	*Spring washer – 2 off*
7	*Ignition coil assembly*	*17*	*Stop lamp switch assembly*	*27*	*Wiring harness*
8	*Suppressor cap*	*18*	*Screw*	*28*	*Oil level gauge assembly*
9	*Ignition switch assembly*	*19*	*Spring washer*	*29*	*Resistor*
10	*Bulb – 3 off*	*20*	*Neutral indicator switch*	*30*	*Grommet*

16 Fault diagnosis: electrical system

Symptom	Cause	Remedy
Complete electrical failure	Blown fuse	Check wiring and electrical components for short circuit before fitting new 10 amp fuse. Check battery connections, also whether connections show signs of corrosion.
Dim lights, horn inoperative	Discharged battery	Re-charge battery with battery charger. Check whether generator is giving correct output.
Constantly blowing bulbs	Vibration, poor earth connection	Check security of bulb holders. Check earth return connections.

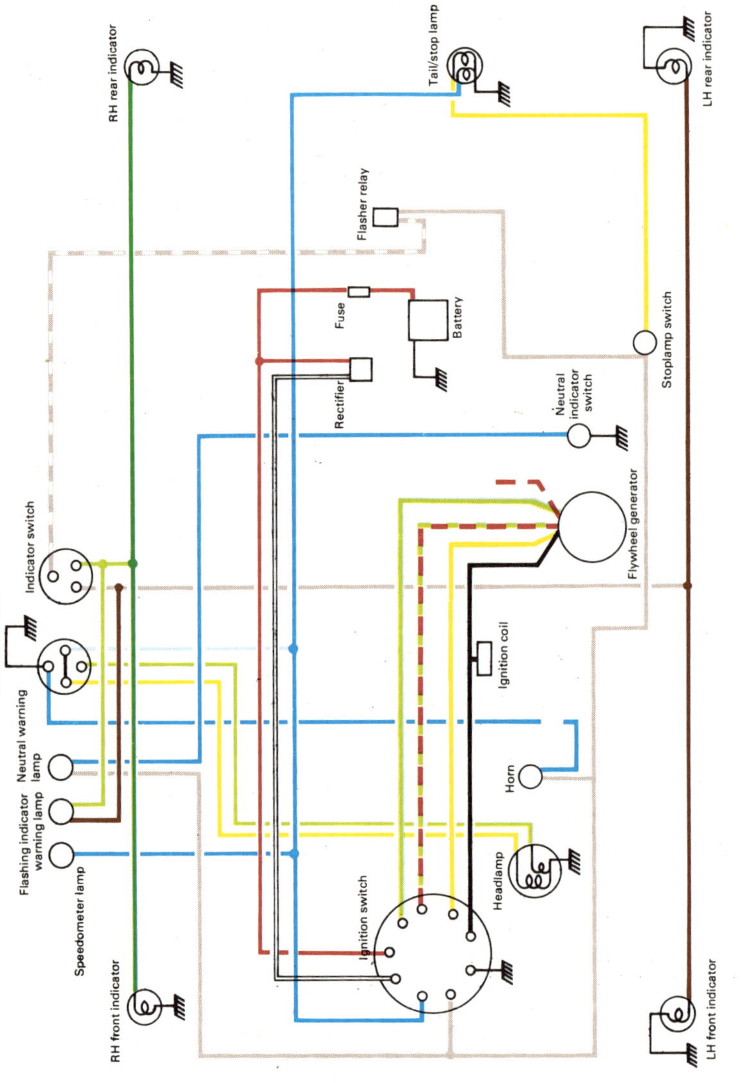

Yamaha YB100 wiring diagram

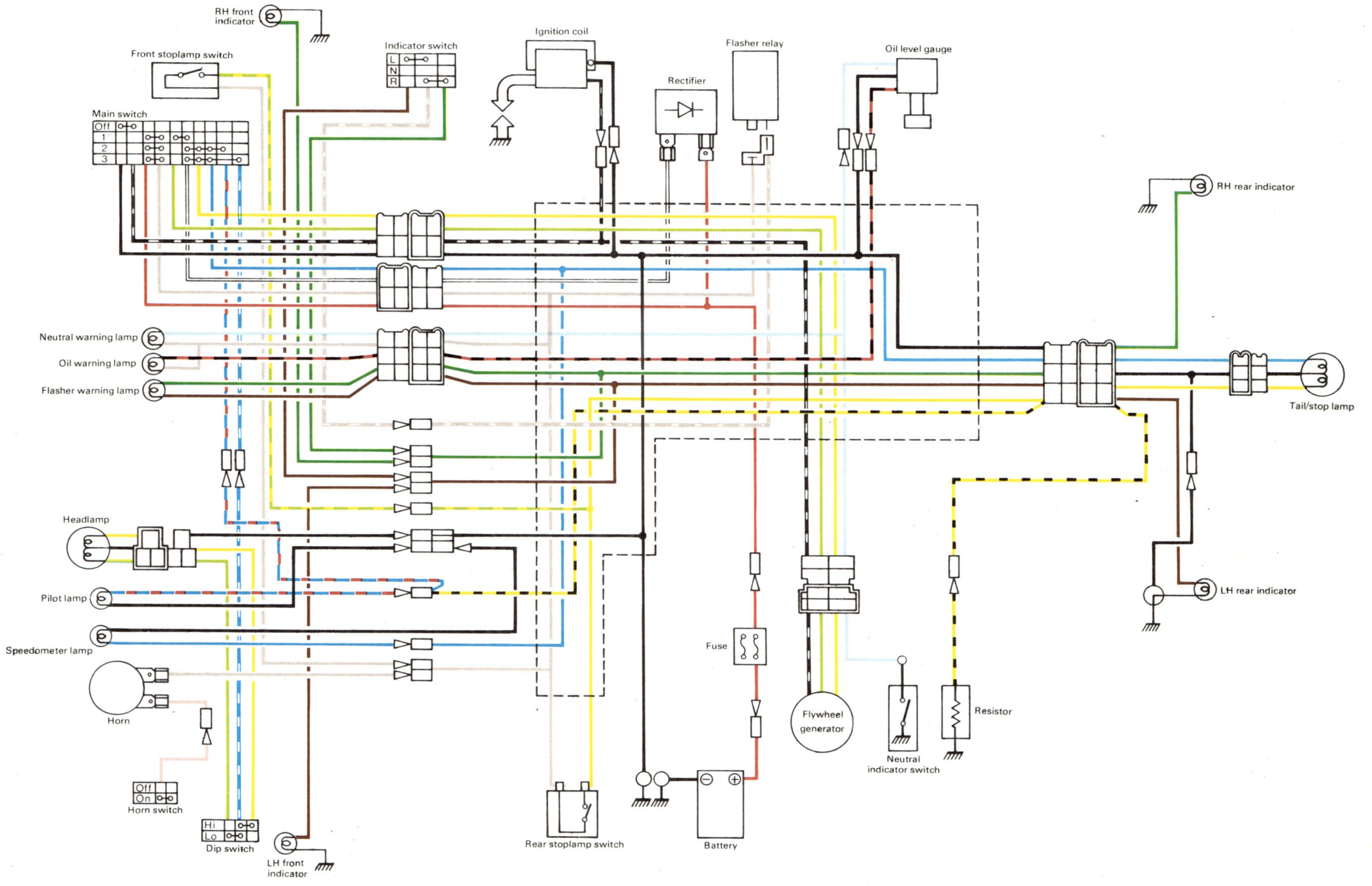

Yamaha YB100E wiring diagram

Metric conversion tables

			Millimetres to Inches		Inches to Millimetres	
Inches	Decimals	Millimetres	mm	Inches	Inches	mm
1/64	0.015625	0.3969	0.01	0.00039	0.001	0.0254
1/32	0.03125	0.7937	0.02	0.00079	0.002	0.0508
3/64	0.046875	1.1906	0.03	0.00118	0.003	0.0762
1/16	0.0625	1.5875	0.04	0.00157	0.004	0.1016
5/64	0.078125	1.9844	0.05	0.00197	0.005	0.1270
3/32	0.09375	2.3812	0.06	0.00236	0.006	0.1524
7/64	0.109375	2.7781	0.07	0.00276	0.007	0.1778
1/8	0.125	3.1750	0.08	0.00315	0.008	0.2032
9/64	0.140625	3.5719	0.09	0.00354	0.009	0.2286
5/32	0.15625	3.9687	0.1	0.00394	0.01	0.254
11/64	0.171875	4.3656	0.2	0.00787	0.02	0.508
3/16	0.1875	4.7625	0.3	0.01181	0.03	0.762
13/64	0.203125	5.1594	0.4	0.01575	0.04	1.016
7/32	0.21875	5.5562	0.5	0.01969	0.05	1.270
15/64	0.234375	5.9531	0.6	0.02362	0.06	1.524
1/4	0.25	6.3500	0.7	0.02756	0.07	1.778
17/64	0.265625	6.7469	0.8	0.03150	0.08	2.032
9/32	0.28125	7.1437	0.9	0.03543	0.09	2.286
19/64	0.296875	7.5406	1	0.03937	0.1	2.54
5/16	0.3125	7.9375	2	0.07874	0.2	5.08
21/64	0.328125	8.3344	3	0.11811	0.3	7.62
11/32	0.34375	8.7312	4	0.15748	0.4	10.16
23/64	0.359375	9.1281	5	0.19685	0.5	12.70
3/8	0.375	9.5250	6	0.23622	0.6	15.24
25/64	0.390625	9.9219	7	0.27559	0.7	17.78
13/32	0.40625	10.3187	8	0.31496	0.8	20.32
27/64	0.421875	10.7156	9	0.35433	0.9	22.86
7/16	0.4375	11.1125	10	0.39370	1	25.4
29/64	0.453125	11.5094	11	0.43307	2	50.8
15/32	0.46875	11.9062	12	0.47244	3	76.2
31/64	0.48375	12.3031	13	0.51181	4	101.6
1/2	0.5	12.7000	14	0.55118	5	127.0
33/64	0.515625	13.0969	15	0.59055	6	152.4
17/32	0.53125	13.4937	16	0.62992	7	177.8
35/64	0.546875	13.8906	17	0.66929	8	203.2
9/16	0.5625	14.2875	18	0.70866	9	228.6
37/64	0.578125	14.6844	19	0.74803	10	254.0
19/32	0.59375	15.0812	20	0.78740	11	279.4
39/64	0.609375	15.4781	21	0.82677	12	304.8
5/8	0.625	15.8750	22	0.86614	13	330.2
41/64	0.640625	16.2719	23	0.90551	14	355.6
21/32	0.65625	16.6687	24	0.94488	15	381.0
43/64	0.671875	17.0656	25	0.98425	16	406.4
11/16	0.6875	17.4625	26	1.02362	17	431.8
45/64	0.703125	17.8594	27	1.06299	18	457.2
23/32	0.71875	18.2562	28	1.10236	19	482.6
47/64	0.734375	18.6531	29	1.14173	20	508.0
3/4	0.75	19.0500	30	1.18110	21	533.4
49/64	0.765625	19.4469	31	1.22047	22	558.8
25/32	0.78125	19.8437	32	1.25984	23	584.2
51/64	0.796875	20.2406	33	1.29921	24	609.6
13/16	0.8125	20.6375	34	1.33858	25	635.0
53/64	0.828125	21.0344	35	1.37795	26	660.4
27/32	0.84375	21.4312	36	1.41732	27	685.8
55/64	0.859375	21.8281	37	1.4567	28	711.2
7/8	0.875	22.2250	38	1.4961	29	736.6
57/64	0.890625	22.6219	39	1.5354	30	762.0
29/32	0.90625	23.0187	40	1.5748	31	787.4
59/64	0.921875	23.4156	41	1.6142	32	812.8
15/16	0.9375	23.8125	42	1.6535	33	838.2
61/64	0.953125	24.2094	43	1.6929	34	863.6
31/32	0.96875	24.6062	44	1.7323	35	889.0
63/64	0.984375	25.0031	45	1.7717	36	914.4

1 Imperial gallon = 8 Imp pints = 1.20 US gallons = 277.42 cu in = 4.54 litres

1 US gallon = 4 US quarts = 0.83 Imp gallon = 231 cu in = 3.78 litres

1 Litre = 0.21 Imp gallon = 0.26 US gallon = 61.02 cu in = 1000 cc

Miles to Kilometres	
1	1.61
2	3.22
3	4.83
4	6.44
5	8.05
6	9.66
7	11.27
8	12.88
9	14.48
10	16.09
20	32.19
30	48.28
40	64.37
50	80.47
60	96.56
70	112.65
80	128.75
90	144.84
100	160.93

Kilometres to Miles	
1	0.62
2	1.24
3	1.86
4	2.49
5	3.11
6	3.73
7	4.35
8	4.97
9	5.59
10	6.21
20	12.43
30	18.64
40	24.85
50	31.07
60	37.28
70	43.50
80	49.71
90	55.92
100	62.14

lbf ft to kgf m	
1	0.138
2	0.276
3	0.414
4	0.553
5	0.691
6	0.829
7	0.967
8	1.106
9	1.244
10	1.382
20	2.765
30	4.147

kgf m to lbf ft	
1	7.233
2	14.466
3	21.699
4	28.932
5	36.165
6	43.398
7	50.631
8	57.864
9	65.097
10	72.330
20	144.660
30	216.990

lbf/in^2 to kgf/cm^2	
1	0.07
2	0.14
3	0.21
4	0.28
5	0.35
6	0.42
7	0.49
8	0.56
9	0.63
10	0.70
20	1.41
30	2.11

kgf/cm^2 to	lbf/in^2
1	14.22
2	28.50
3	42.67
4	56.89
5	71.12
6	85.34
7	99.56
8	113.79
9	128.00
10	142.23
20	284.47
30	426.70

Index

A
Acknowledgements 2
About this manual 2
Adjustments:-
    brakes 91
    carburettor 62
    clutch 10
    contact breaker 10, 68
    final drive chain 93
    front brake 91
    headlamp beam height 99
    horn 106
    oil pump 10
    rear brake 91
    slow running speed 10
Air filter 62

B

Battery charging output 78
Bearings:-
    big-end 31
    main 30
    small-end 31
    steering head 13, 75
    wheel 13, 89
Brakes:-
    dismantling, examination and replacement 88
    front 86
    pedal – rear brake 82
Bulbs – replacement:-
    flasher 103
    headlamp 99
    speedometer 103
    stop and tail 103
Buying:-
    spare parts 7
    tools 15

C

Cables:-
    clutch 10
    brake 91
    lubrication 9
    pump control 21, 50
    speedometer drive 83
    throttle 50
Carburettor:-
    adjustment 62
    dismantling and reassembly 58
    fault diagnosis 66
    removal 57
    settings 59
    specifications 55
Chain – final drive 93
Chapter contents:-
    1 Engine, clutch and gearbox 16
    2 Fuel system and lubrication 55
    3 Ignition system 69
    4 Frame and forks 73
    5 Wheels, brakes and tyres 84
    6 Electrical system 97
Checks:-
    battery 9
    brake linings 13
    carburettor settings 59
    coil 71
    contact breakers 10, 68
    electrical system 48
    legal 9
    rectifier 98
    safety 9
    sparking plug gap settings 72
    tyre pressures 9, 84
Centre stand 81
Cleaning:-
    air filter 11, 62
    exhaust systems 13
    final drive chain 12
    fuel tap filter 10
    the machine 83
Clutch:-
    adjustment 10
    cable 10
    drag 54
    examination and renovation 36
    fault diagnosis
    removal 25
    refitting 45
    slip 36, 54
    specifications 17
Coil – ignition 71
Condenser 70
Contact breaker 68, 70
Crankcase halves:-
    separating 29
    joining 40
Crankshaft 31
Cush drive 93
Cylinder barrel 21, 33
Cylinder head 21, 33

D

Decarbonising:-
    engine 13, 33
    silencer 64
Description – general:-
    electrical system 98
    engine, clutch and gearbox 17
    frame and forks 93
    fuel system 55
    ignition system 68
    lubrication 55
    wheels, brakes and tyres 84
Disc valve 62
Dimensions and weight 6
Dust caps – tyre valves 96

E

Electrical system:-
    battery 99
    fault diagnosis 109
    flashing indicators 103

flywheel generator 29, 48
fuse location 99
headlamps 99
horn 106
lamps 103
rectifier 98
specifications 97, 98
switches – handlebar 102
wiring diagrams – colour 110, 111
Engine:-
bearings:
big-end 30
main 31
clutch removal 25
connecting rod 21
crankshaft 31
crankcase halves:
separating 29
joining 40
cylinder barrel 21, 33
cylinder head 21, 33
decarbonising:
engine 13
silencer 64
disc valve 29, 40, 62
dismantling–general 21
examination and renovation – general 30
fault diagnosis 53
gudgeon pin 21, 31, 33
lubrication system 56, 62
oil seals 30
oil pump:
removal 21
refitting 64
piston and rings 33
reassembly general 37
specifications 16, 17
starting and running a rebuilt unit 53
Exhaust system 64

F

Fault diagnosis:-
clutch 54
electrical system 109
engine 53
frame and forks 83
fuel system 66
gearbox 54
ignition system 72
lubrication 66
wheels, brakes and tyres 96
Filter – air 62
Final drive chain 93
Flywheel generator 29, 48
Footrests 81
Frame and forks:-
centre stand 81
fault diagnosis 83
foot rests 81
fork yokes 75
frame examination and renovation 79
front fork legs:
dismantling 75
examination and renovation 76
removal 74
reassembly and replacement 78
kickstart lever 82
prop stand 81
rear brake pedal 82
rear suspension units 80
speedometer drive 83
steering head lock 78
swinging arm rear fork 79
Front wheel 84
Fuel system:-
air filter 62
carburettor:
adjustment 62
dismantling 58
removal 57
settings 59
fault diagnosis 66
petrol tank 56
petrol tap 57
Fuel location 99

G

Gearbox:-
components:
removal 30
renovation 34
dismantling – general 21
fault diagnosis 54
gearchange mechanism removal 26
removal 18
replacing gearchange mechanism 42
specifications 17
Generator – flywheel 29, 44
Gudgeon pin 21, 31, 33

H

Handlebar switches 102
Headlamp beam height adjustment 99
Horn location 106

I

Ignition system:-
coil checking 71
condenser 70
contact breaker 68, 70
fault diagnosis 68
flywheel generator 29, 48
sparking plug:
checking and setting gap 72
operating conditions – colour chart 68
specifications 68
timing 68

K

Kickstart:-
lever examination and renovation 82
removal 36
refitting 43

L

Lamps 99, 103
Legal obligations 9, 96, 99
Lubrication:-
control cables 4, 11
final drive chain 93
general 11
wheel bearings 13
Lubricants – recommended 14

M

Maintenance – adjustments and capacities 14
Maintenance – routine 8-13
Metric conversion tables 112-113

O

Oil pump:-
  bleeding 64
  control cable 21-50
  removal 21-64
  replacement 64
Ordering:-
  spare parts 7
  tools 15
Oil seals 30

P

Pedal – rear brake 82
Petrol tank 56
Petrol tap 57
Piston and rings 33
Prop stand 81

Q

Quick glance adjustments and capacities 14

R

Rear brake pedal 82
Rear chain:-
  cleaning and lubrication 12, 93
  examination 93
Rear suspension rim units 80
Rear wheel:-
  cush drive 97
  sprockets 93
Rectifier 98
Rings and piston 33
Routine maintenance 8-13

S

Safety check 9
Sparking plugs:-
  checking and setting gap 72
  operating conditions – colour chart 67
Speedometer drive 85
Specifications:-
  bulbs 98
  clutch 17
  electrical system 97, 98
  engine 16-17
  frame and forks 73
  fuel system 55
  gearbox 17
  ignition 68
  lubrication 55
  wheels, brakes and tyres 84
Statutory requirements 9, 96, 99
Suspension unit – rear 80
Swinging arm rear fork 79
Switches:-
  brake light 103
  ignition 102
  lighting 102

T

Tools 15
Torque wrench settings 14
Tyres:-
  pressures 9, 84
  removal and replacement – colour instructions 95
  valves and dust caps 96

V

Valve – disc 29, 40, 62

W

Weight and dimensions 8
Wheels:-
  bearings 13, 89
  front 89
  rear 56
Wiring diagrams – colour 110, 111
Working conditions 15

BEDFORDSHIRE COUNTY LIBRARY COUNTY COUNCIL

Printed by
Haynes Publishing Group
Sparkford Yeovil Somerset
England